T0296878

Lecture Notes in Computational Science and Engineering

110

Editors:

More information about this series at http://www.springer.com/series/3527

Hans Petter Langtangen

Finite Difference Computing with Exponential Decay Models

Springer Open

Hans Petter Langtangen
Simula Research Laboratory
Lysaker, Norway

On leave from:

Department of Informatics
University of Oslo
Oslo, Norway

ISSN 1439-7358 ISSN 2197-7100 (electronic)
Lecture Notes in Computational Science and Engineering
ISBN 978-3-319-29438-4 ISBN 978-3-319-29439-1 (eBook)
DOI 10.1007/978-3-319-29439-1
Springer Cham Heidelberg New York Dordrecht London

Library of Congress Control Number: 2016932614

Mathematic Subject Classification (2010): 34, 65, 68

Printed on acid-free paper

Springer International Publishing AG Switzerland is part of Springer Science+Business Media
(www.springer.com)

Preface

This book teaches the basic components in the scientific computing pipeline: modeling, differential equations, numerical algorithms, programming, plotting, and software testing. The pedagogical idea is to treat these topics in the context of a very simple mathematical model, the differential equation for exponential decay, $u'(t) = -au(t)$, where u is unknown and a is a given parameter. By keeping the mathematical problem simple, the text can go deep into all details about how one must combine mathematics and computer science to create well-tested, reliable, and flexible software for such a mathematical model.

The writing style is gentle and aims at a broad audience. I am much inspired by Nick Trefethen's praise of easy learning:

> Some people think that stiff challenges are the best device to induce learning, but I am not one of them. The natural way to learn something is by spending vast amounts of easy, enjoyable time at it. This goes whether you want to speak German, sight-read at the piano, type, or do mathematics. Give me the German storybook for fifth graders that I feel like reading in bed, not Goethe and a dictionary. The latter will bring rapid progress at first, then exhaustion and failure to resolve.
>
> The main thing to be said for stiff challenges is that inevitably we will encounter them, so we had better learn to face them boldly. Putting them in the curriculum can help teach us to do so. But for teaching the skill or subject matter itself, they are overrated. [13, p. 86]

Prerequisite knowledge for this book is basic one-dimensional calculus and preferably some experience with computer programming in Python or MATLAB. The material was initially written for self study and therefore features comprehensive and easy-to-understand explanations. For some readers it may act as an overview and refresher of traditional mathematical topics and likely a first introduction to many of the software topics. The text can also be used as a case-based and mathematically simple introduction to modern multi-disciplinary problem solving with computers, using the range of applications in Chap. 4 as motivation and then treating the details of the mathematical and computer science subjects from the other chapters. In particular, I have also had in mind the new groups of readers from bio- and geo-sciences who need to enter the world of computer-based differential equation modeling, but lack experience with (and perhaps also interest in) mathematics and programming.

The choice of topics in this book is motivated from what is needed in more advanced courses on finite difference methods for partial differential equations

(PDEs). It turns out that a range of concepts and tools needed for PDEs can be introduced and illustrated by very simple ordinary differential equation (ODE) examples. The goal of the text is therefore to lay a foundation for understanding numerical methods for PDEs by first meeting the fundamental ideas in a simpler ODE setting. Compared to other books, the present one has a much stronger focus on how to turn mathematics into working code. It also explains the mathematics and programming in more detail than what is common in the literature.

There is a more advanced companion book in the works, "Finite Difference Computing with Partial Differential Equations", which treats finite difference methods for PDEs using the same writing style and having the same focus on turning mathematical algorithms into reliable software.

Although the main example in the present book is $u' = -au$, we also address the more general model problem $u' = -a(t)u + b(t)$, and the completely general, nonlinear problem $u' = f(u,t)$, both for scalar and vector $u(t)$. The author believes in the principle *simplify, understand, and then generalize*. That is why we start out with the simple model $u' = -au$ and try to understand how methods are constructed, how they work, how they are implemented, and how they may fail for this problem, before we generalize what we have learned from $u' = -au$ to more complicated models.

The following list of topics will be elaborated on.

- How to think when constructing finite difference methods, with special focus on the Forward Euler, Backward Euler, and Crank–Nicolson (midpoint) schemes.
- How to formulate a computational algorithm and translate it into Python code.
- How to make curve plots of the solutions.
- How to compute numerical errors.
- How to compute convergence rates.
- How to test that an implementation is correct (verification) and how to automate tests through *test functions* and *unit testing*.
- How to work with Python concepts such as arrays, lists, dictionaries, lambda functions, and functions in functions (closures).
- How to perform array computing and understand the difference from scalar computing.
- How to uncover numerical artifacts in the computed solution.
- How to analyze the numerical schemes mathematically to understand why artifacts may occur.
- How to derive mathematical expressions for various measures of the error in numerical methods, frequently by using the `sympy` software for symbolic computations.
- How to understand concepts such as finite difference operators, mesh (grid), mesh functions, stability, truncation error, consistency, and convergence.
- How to solve the general nonlinear ODE $u' = f(u,t)$, which is either a scalar ODE or a system of ODEs (i.e., u and f can either be a function or a vector of functions).
- How to access professional packages for solving ODEs.
- How the model equation $u' = -au$ arises in a wide range of phenomena in physics, biology, chemistry, and finance.
- How to structure a code in terms of functions.

- How to make reusable modules.
- How to read input data flexibly from the command line.
- How to create graphical/web user interfaces.
- How to use test frameworks for automatic unit testing.
- How to refactor code in terms of classes (instead of functions).
- How to conduct and automate large-scale numerical experiments.
- How to write scientific reports in various formats (LaTeX, HTML).

The exposition in a nutshell

Everything we cover is put into a practical, hands-on context. All mathematics is translated into working computing codes, and all the mathematical theory of finite difference methods presented here is motivated from a strong need to understand why we occasionally obtain strange results from the programs. Two fundamental questions saturate the text:

- How do we solve a differential equation problem and produce numbers?
- How do we know that the numbers are correct?

Besides answering these two questions, one will learn a lot about mathematical modeling in general and the interplay between physics, mathematics, numerical methods, and computer science.

The book contains a set of exercises in most of the chapters. The exercises are divided into three categories: *exercises* refer to the text (usually variations or extensions of examples in the text), *problems* are stand-alone exercises without references to the text, and *projects* are larger problems. Exercises, problems, and projects share a common numbering to avoid confusion between, e.g., Exercise 4.3 and Problem 4.3 (it will be Exercise 4.3 and Problem 4.4 if they follow after each other).

All program and data files referred to in this book are available from the book's primary web site: http://hplgit.github.io/decay-book/doc/web/.

Acknowledgments Professor Svein Linge provided very detailed and constructive feedback on this text, and all his efforts are highly appreciated. Many students have also pointed out weaknesses and found errors. A special thank goes to Yapi Donatien Achou's proof reading. Many thanks also to Linda Falch-Koslung, Dr. Olav Dajani, and the rest of the OUS team for feeding me with FOLFIRINOX and thereby keeping me alive and in good enough shape to finish this book. As always, the Springer team ensured a smooth and rapid review process and production phase. This time special thanks go to all the efforts by Martin Peters, Thanh-Ha Le Thi, and Yvonne Schlatter.

Oslo, August 2015 Hans Petter Langtangen

Contents

List of Exercises, Problems, and Projects

Algorithms and Implementations

1

Throughout industry and science it is common today to study nature or technological devices through models on a computer. With such models the computer acts as a virtual lab where experiments can be done in a fast, reliable, safe, and cheap way. In some fields, e.g., aerospace engineering, the computer models are now so sophisticated that they can replace physical experiments to a large extent.

A vast amount of computer models are based on ordinary and partial differential equations. This book is an introduction to the various scientific ingredients we need for reliable computing with such type of models. A key theme is to solve differential equations *numerically* on a computer. Many methods are available for this purpose, but the focus here is on *finite difference methods*, because these are simple, yet versatile, for solving a wide range of ordinary and partial differential equations. The present chapter first presents the mathematical ideas of finite difference methods and derives algorithms, i.e., formulations of the methods ready for computer programming. Then we create programs and learn how we can be sure that the programs really work correctly.

1.1 Finite Difference Methods

This section explains the basic ideas of finite difference methods via the simple ordinary differential equation $u' = -au$. Emphasis is put on the reasoning around discretization principles and introduction of key concepts such as mesh, mesh function, finite difference approximations, averaging in a mesh, derivation of algorithms, and discrete operator notation.

1.1.1 A Basic Model for Exponential Decay

Our model problem is perhaps the simplest ordinary differential equation (ODE):

$$u'(t) = -au(t) .$$

In this equation, $u(t)$ is a scalar function of time t, a is a constant (in this book we mostly work with $a > 0$), and $u'(t)$ means differentiation with respect to t.

H.P. Langtangen, *Finite Difference Computing with Exponential Decay Models*,
Lecture Notes in Computational Science and Engineering 110,
DOI 10.1007/978-3-319-29439-1_1

This type of equation arises in a number of widely different phenomena where some quantity u undergoes exponential reduction (provided $a > 0$). Examples include radioactive decay, population decay, investment decay, cooling of an object, pressure decay in the atmosphere, and retarded motion in fluids. Some models with growth, $a < 0$, are treated as well, see Chap. 4 for details and motivation. We have chosen this particular ODE not only because its applications are relevant, but even more because studying numerical solution methods for this particular ODE gives important insight that can be reused in far more complicated settings, in particular when solving diffusion-type partial differential equations.

The exact solution Although our interest is in *approximate* numerical solutions of $u' = -au$, it is convenient to know the exact analytical solution of the problem so we can compute the error in numerical approximations. The analytical solution of this ODE is found by separation of variables, which results in

$$u(t) = Ce^{-at},$$

for any arbitrary constant C. To obtain a unique solution, we need a condition to fix the value of C. This condition is known as the *initial condition* and stated as $u(0) = I$. That is, we know that the value of u is I when the process starts at $t = 0$. With this knowledge, the exact solution becomes $u(t) = Ie^{-at}$. The initial condition is also crucial for numerical methods: without it, we can never start the numerical algorithms!

A complete problem formulation Besides an initial condition for the ODE, we also need to specify a time interval for the solution: $t \in (0, T]$. The point $t = 0$ is not included since we know that $u(0) = I$ and assume that the equation governs u for $t > 0$. Let us now summarize the information that is required to state the complete problem formulation: find $u(t)$ such that

$$u' = -au,\ t \in (0, T], \quad u(0) = I\,. \tag{1.1}$$

This is known as a *continuous problem* because the parameter t varies continuously from 0 to T. For each t we have a corresponding $u(t)$. There are hence infinitely many values of t and $u(t)$. The purpose of a numerical method is to formulate a corresponding *discrete* problem whose solution is characterized by a finite number of values, which can be computed in a finite number of steps on a computer. Typically, we choose a finite set of time values $t_0, t_1, \ldots, t_{N_t}$, and create algorithms that generate the corresponding u values $u_0, u_1, \ldots, u_{N_t}$.

1.1.2 The Forward Euler Scheme

Solving an ODE like (1.1) by a finite difference method consists of the following four steps:

1. discretizing the domain,
2. requiring fulfillment of the equation at discrete time points,
3. replacing derivatives by finite differences,
4. formulating a recursive algorithm.

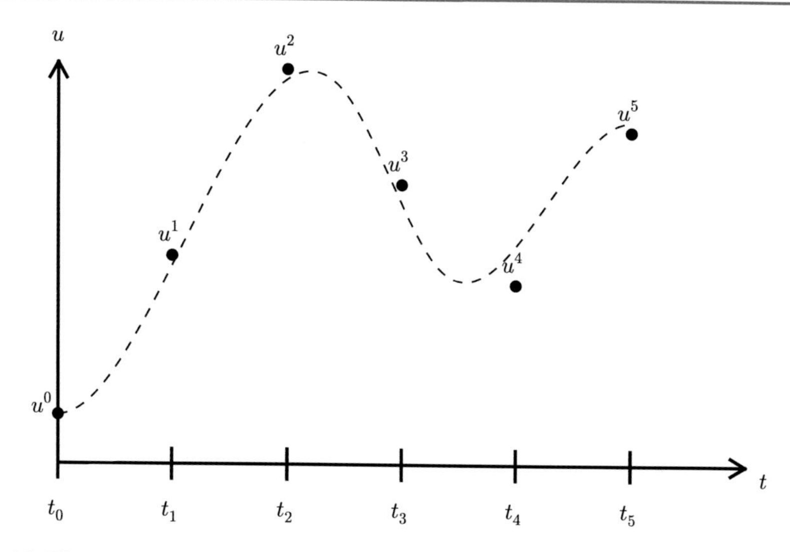

Fig. 1.1 Time mesh with discrete solution values at *points* and a *dashed line* indicating the true solution

Step 1: Discretizing the domain The time domain $[0, T]$ is represented by a finite number of $N_t + 1$ points

$$0 = t_0 < t_1 < t_2 < \cdots < t_{N_t-1} < t_{N_t} = T\,. \tag{1.2}$$

The collection of points $t_0, t_1, \ldots, t_{N_t}$ constitutes a *mesh* or *grid*. Often the mesh points will be uniformly spaced in the domain $[0, T]$, which means that the spacing $t_{n+1} - t_n$ is the same for all n. This spacing is often denoted by Δt, which means that $t_n = n\Delta t$.

We want the solution u at the mesh points: $u(t_n)$, $n = 0, 1, \ldots, N_t$. A notational short-form for $u(t_n)$, which will be used extensively, is u^n. More precisely, we let u^n be the *numerical approximation* to the exact solution $u(t_n)$ at $t = t_n$.

When we need to clearly distinguish between the numerical and exact solution, we often place a subscript e on the exact solution, as in $u_{\mathrm{e}}(t_n)$. Figure 1.1 shows the t_n and u^n points for $n = 0, 1, \ldots, N_t = 7$ as well as $u_{\mathrm{e}}(t)$ as the dashed line.

We say that the numerical approximation, i.e., the collection of u^n values for $n = 0, \ldots, N_t$, constitutes a *mesh function*. A "normal" continuous function is a curve defined for all real t values in $[0, T]$, but a mesh function is only defined at discrete points in time. If you want to compute the mesh function *between* the mesh points, where it is not defined, an *interpolation method* must be used. Usually, linear interpolation, i.e., drawing a straight line between the mesh function values, see Fig. 1.1, suffices. To compute the solution for some $t \in [t_n, t_{n+1}]$, we use the linear interpolation formula

$$u(t) \approx u^n + \frac{u^{n+1} - u^n}{t_{n+1} - t_n}(t - t_n)\,. \tag{1.3}$$

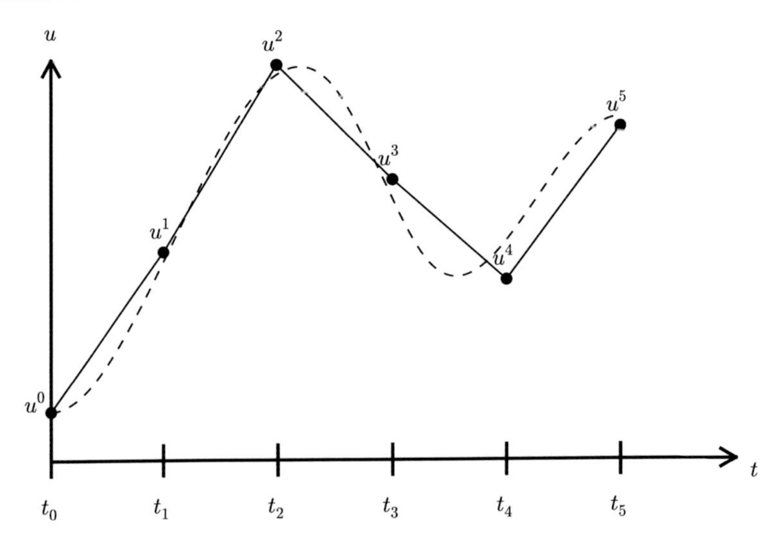

Fig. 1.2 Linear interpolation between the discrete solution values (*dashed curve* is exact solution)

Notice

The goal of a numerical solution method for ODEs is to compute the mesh function by solving a finite set of *algebraic equations* derived from the original ODE problem.

Step 2: Fulfilling the equation at discrete time points The ODE is supposed to hold for all $t \in (0, T]$, i.e., at an infinite number of points. Now we relax that requirement and require that the ODE is fulfilled at a finite set of discrete points in time. The mesh points $t_0, t_1, \ldots, t_{N_t}$ are a natural (but not the only) choice of points. The original ODE is then reduced to the following equations:

$$u'(t_n) = -au(t_n), \quad n = 0, \ldots, N_t, \quad u(0) = I\,. \tag{1.4}$$

Even though the original ODE is not stated to be valid at $t = 0$, it is valid as close to $t = 0$ as we like, and it turns out that it is useful for construction of numerical methods to have (1.4) valid for $n = 0$. The next two steps show that we need (1.4) for $n = 0$.

Step 3: Replacing derivatives by finite differences The next and most essential step of the method is to replace the derivative u' by a finite difference approximation. Let us first try a *forward* difference approximation (see Fig. 1.3),

$$u'(t_n) \approx \frac{u^{n+1} - u^n}{t_{n+1} - t_n}\,. \tag{1.5}$$

The name forward relates to the fact that we use a value forward in time, u^{n+1}, together with the value u^n at the point t_n, where we seek the derivative, to approximate

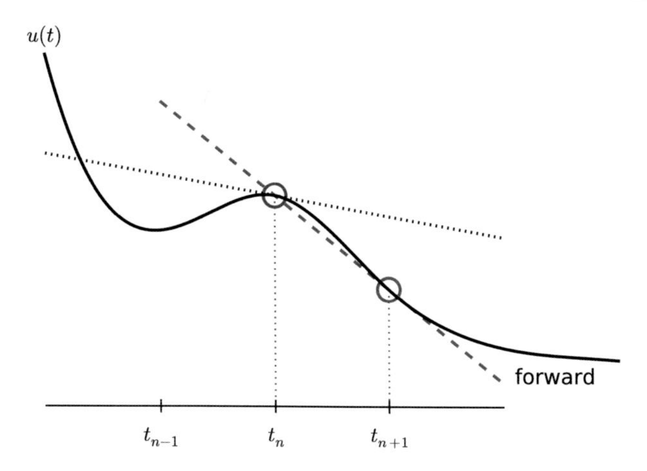

Fig. 1.3 Illustration of a forward difference

$u'(t_n)$. Inserting this approximation in (1.4) results in

$$\frac{u^{n+1} - u^n}{t_{n+1} - t_n} = -au^n, \quad n = 0, 1, \ldots, N_t - 1\,. \tag{1.6}$$

Note that if we want to compute the solution up to time level N_t, we only need (1.4) to hold for $n = 0, \ldots, N_t - 1$ since (1.6) for $n = N_t - 1$ creates an equation for the final value u^{N_t}.

Also note that we use the approximation symbol $\approx$ in (1.5), but not in (1.6). Instead, we view (1.6) as an equation that is not mathematically equivalent to (1.5), but represents an approximation to (1.5).

Equation (1.6) is the discrete counterpart to the original ODE problem (1.1), and often referred to as a *finite difference scheme* or more generally as the *discrete equations* of the problem. The fundamental feature of these equations is that they are *algebraic* and can hence be straightforwardly solved to produce the mesh function, i.e., the approximate values of u at the mesh points: u^n, $n = 1, 2, \ldots, N_t$.

Step 4: Formulating a recursive algorithm The final step is to identify the computational algorithm to be implemented in a program. The key observation here is to realize that (1.6) can be used to compute u^{n+1} if u^n is known. Starting with $n = 0$, u^0 is known since $u^0 = u(0) = I$, and (1.6) gives an equation for u^1. Knowing u^1, u^2 can be found from (1.6). In general, u^n in (1.6) can be assumed known, and then we can easily solve for the unknown u^{n+1}:

$$u^{n+1} = u^n - a(t_{n+1} - t_n)u^n\,. \tag{1.7}$$

We shall refer to (1.7) as the Forward Euler (FE) scheme for our model problem. From a mathematical point of view, equations of the form (1.7) are known as *difference equations* since they express how differences in the dependent variable, here u, evolve with n. In our case, the differences in u are given by $u^{n+1} - u^n =$

$-a(t_{n+1} - t_n)u^n$. The finite difference method can be viewed as a method for turning a differential equation into an algebraic difference equation that can be easily solved by repeated use of a formula like (1.7).

Interpretation There is a very intuitive interpretation of the FE scheme, illustrated in the sketch below. We have computed some point values on the solution curve (small red disks), and the question is how we reason about the next point. Since we know u and t at the most recently computed point, the differential equation gives us the *slope* of the solution curve: $u' = -au$. We can draw this slope as a red line and continue the solution curve along that slope. As soon as we have chosen the next point on this line, we have a new t and u value and can compute a new slope and continue the process.

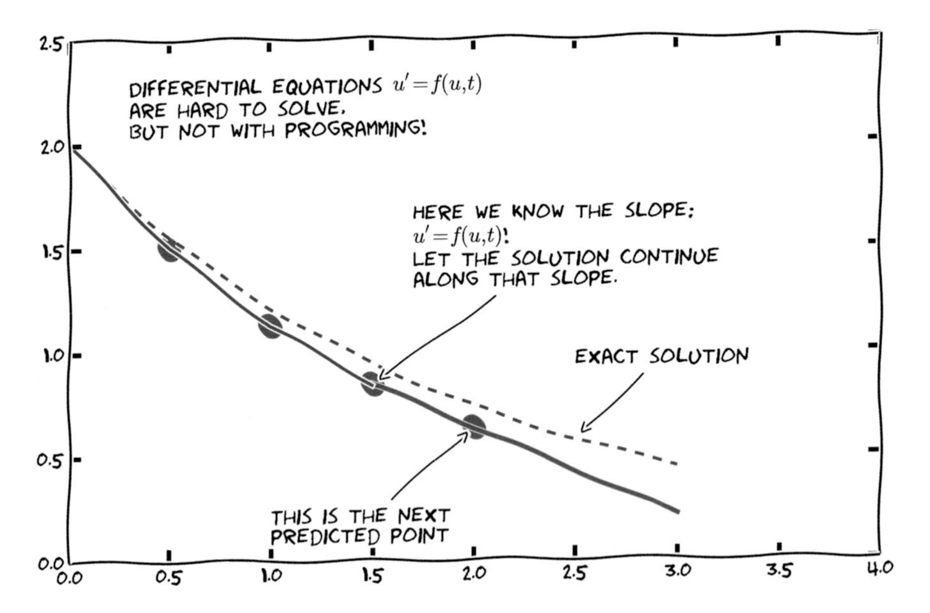

Computing with the recursive formula Mathematical computation with (1.7) is straightforward:

$$
\begin{aligned}
u_0 &= I,\\
u_1 &= u^0 - a(t_1 - t_0)u^0 = I(1 - a(t_1 - t_0)),\\
u_2 &= u^1 - a(t_2 - t_1)u^1 = I(1 - a(t_1 - t_0))(1 - a(t_2 - t_1)),\\
u^3 &= u^2 - a(t_3 - t_2)u^2 = I(1 - a(t_1 - t_0))(1 - a(t_2 - t_1))(1 - a(t_3 - t_2)),
\end{aligned}
$$

and so on until we reach u^{N_t}. Very often, $t_{n+1} - t_n$ is constant for all n, so we can introduce the common symbol $\Delta t = t_{n+1} - t_n$, $n = 0, 1, \ldots, N_t - 1$. Using a constant mesh spacing Δt in the above calculations gives

$$
\begin{aligned}
u_0 &= I,\\
u_1 &= I(1 - a\Delta t),
\end{aligned}
$$

$$
\begin{aligned}
u_2 &= I(1 - a\Delta t)^2, \\
u^3 &= I(1 - a\Delta t)^3, \\
&\vdots \\
u^{N_t} &= I(1 - a\Delta t)^{N_t}.
\end{aligned}
$$

This means that we have found a closed formula for u^n, and there is no need to let a computer generate the sequence $u^1, u^2, u^3, \ldots$ However, finding such a formula for u^n is possible only for a few very simple problems, so in general finite difference equations must be solved on a computer.

As the next sections will show, the scheme (1.7) is just one out of many alternative finite difference (and other) methods for the model problem (1.1).

1.1.3 The Backward Euler Scheme

There are several choices of difference approximations in step 3 of the finite difference method as presented in the previous section. Another alternative is

$$
u'(t_n) \approx \frac{u^n - u^{n-1}}{t_n - t_{n-1}}. \tag{1.8}
$$

Since this difference is based on going backward in time (t_{n-1}) for information, it is known as a *backward* difference, also called Backward Euler difference. Figure 1.4 explains the idea.

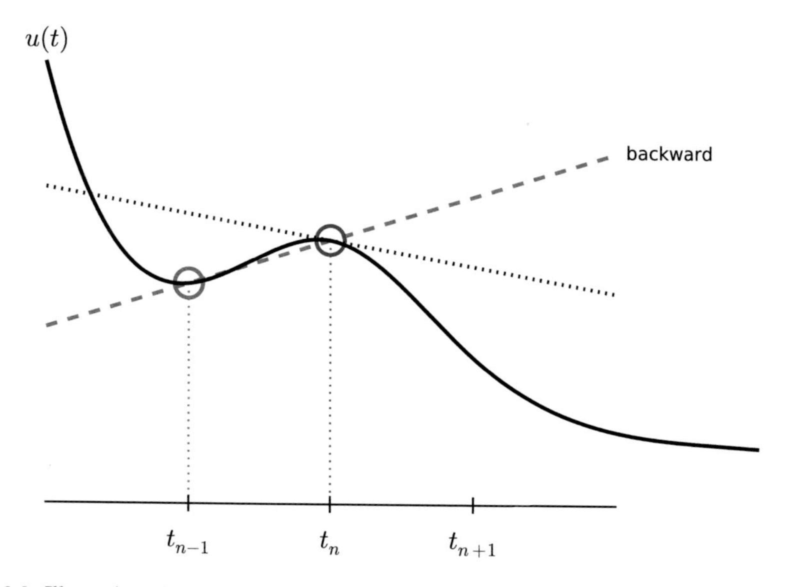

Fig. 1.4 Illustration of a backward difference

Inserting (1.8) in (1.4) yields the Backward Euler (BE) scheme:

$$\frac{u^n - u^{n-1}}{t_n - t_{n-1}} = -au^n, \quad n = 1, \ldots, N_t . \tag{1.9}$$

We assume, as explained under step 4 in Sect. 1.1.2, that we have computed $u^0, u^1, \ldots, u^{n-1}$ such that (1.9) can be used to compute u^n. Note that (1.9) needs n to start at 1 (then it involves u^0, but no u^{-1}) and end at N_t.

For direct similarity with the formula for the Forward Euler scheme (1.7) we replace n by $n + 1$ in (1.9) and solve for the unknown value u^{n+1}:

$$u^{n+1} = \frac{1}{1 + a(t_{n+1} - t_n)} u^n, \quad n = 0, \ldots, N_t - 1 . \tag{1.10}$$

1.1.4 The Crank–Nicolson Scheme

The finite difference approximations (1.5) and (1.8) used to derive the schemes (1.7) and (1.10), respectively, are both one-sided differences, i.e., we collect information either forward or backward in time when approximating the derivative at a point. Such one-sided differences are known to be less accurate than central (or midpoint) differences, where we use information both forward and backward in time. A natural next step is therefore to construct a central difference approximation that will yield a more accurate numerical solution.

The central difference approximation to the derivative is sought at the point $t_{n+\frac{1}{2}} = \frac{1}{2}(t_n + t_{n+1})$ (or $t_{n+\frac{1}{2}} = (n + \frac{1}{2})\Delta t$ if the mesh spacing is uniform in time). The approximation reads

$$u'(t_{n+\frac{1}{2}}) \approx \frac{u^{n+1} - u^n}{t_{n+1} - t_n} . \tag{1.11}$$

Figure 1.5 sketches the geometric interpretation of such a centered difference. Note that the fraction on the right-hand side is the same as for the Forward Euler approximation (1.5) and the Backward Euler approximation (1.8) (with n replaced by $n + 1$). The accuracy of this fraction as an approximation to the derivative of u depends on *where* we seek the derivative: in the center of the interval $[t_n, t_{n+1}]$ or at the end points. We shall later see that it is more accurate at the center point.

With the formula (1.11), where u' is evaluated at $t_{n+\frac{1}{2}}$, it is natural to demand the ODE to be fulfilled at the time points *between* the mesh points:

$$u'(t_{n+\frac{1}{2}}) = -au(t_{n+\frac{1}{2}}), \quad n = 0, \ldots, N_t - 1 . \tag{1.12}$$

Using (1.11) in (1.12) results in the approximate discrete equation

$$\frac{u^{n+1} - u^n}{t_{n+1} - t_n} = -au^{n+\frac{1}{2}}, \quad n = 0, \ldots, N_t - 1, \tag{1.13}$$

where $u^{n+\frac{1}{2}}$ is a short form for the numerical approximation to $u(t_{n+\frac{1}{2}})$.

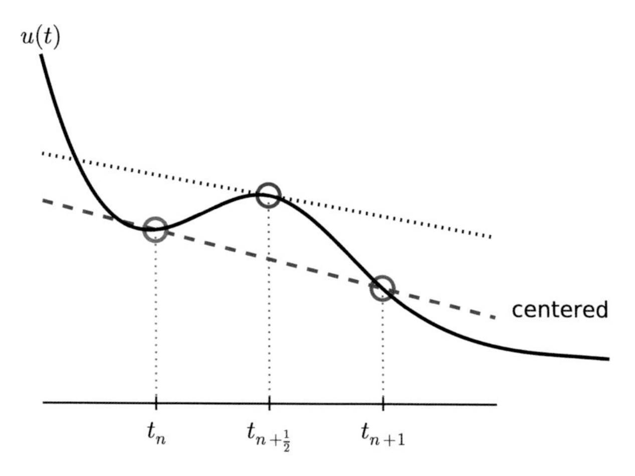

Fig. 1.5 Illustration of a centered difference

There is a fundamental problem with the right-hand side of (1.13): we aim to compute u^n for integer n, which means that $u^{n+\frac{1}{2}}$ is not a quantity computed by our method. The quantity must therefore be expressed by the quantities that we actually produce, i.e., the numerical solution at the mesh points. One possibility is to approximate $u^{n+\frac{1}{2}}$ as an arithmetic mean of the u values at the neighboring mesh points:

$$u^{n+\frac{1}{2}} \approx \frac{1}{2}(u^n + u^{n+1})\,. \tag{1.14}$$

Using (1.14) in (1.13) results in a new approximate discrete equation

$$\frac{u^{n+1} - u^n}{t_{n+1} - t_n} = -a\frac{1}{2}(u^n + u^{n+1})\,. \tag{1.15}$$

There are three approximation steps leading to this formula: 1) the ODE is only valid at discrete points (between the mesh points), 2) the derivative is approximated by a finite difference, and 3) the value of u between mesh points is approximated by an arithmetic mean value. Despite one more approximation than for the Backward and Forward Euler schemes, the use of a centered difference leads to a more accurate method.

To formulate a recursive algorithm, we assume that u^n is already computed so that u^{n+1} is the unknown, which we can solve for:

$$u^{n+1} = \frac{1 - \frac{1}{2}a(t_{n+1} - t_n)}{1 + \frac{1}{2}a(t_{n+1} - t_n)}u^n\,. \tag{1.16}$$

The finite difference scheme (1.16) is often called the Crank–Nicolson (CN) scheme or a midpoint or centered scheme. Note that (1.16) as well as (1.7) and (1.10) apply whether the spacing in the time mesh, $t_{n+1} - t_n$, depends on n or is constant.

1.1.5 The Unifying θ-Rule

The Forward Euler, Backward Euler, and Crank–Nicolson schemes can be formulated as one scheme with a varying parameter θ:

$$\frac{u^{n+1}-u^n}{t_{n+1}-t_n} = -a(\theta u^{n+1} + (1-\theta)u^n)\,. \tag{1.17}$$

Observe that

- $\theta = 0$ gives the Forward Euler scheme
- $\theta = 1$ gives the Backward Euler scheme,
- $\theta = \frac{1}{2}$ gives the Crank–Nicolson scheme.

We shall later, in Chap. 2, learn the pros and cons of the three alternatives. One may alternatively choose any other value of θ in $[0, 1]$, but this is not so common since the accuracy and stability of the scheme do not improve compared to the values $\theta = 0, 1, \frac{1}{2}$.

As before, u^n is considered known and u^{n+1} unknown, so we solve for the latter:

$$u^{n+1} = \frac{1-(1-\theta)a(t_{n+1}-t_n)}{1+\theta a(t_{n+1}-t_n)}\,. \tag{1.18}$$

This scheme is known as the θ-rule, or alternatively written as the "theta-rule".

Derivation

We start with replacing u' by the fraction

$$\frac{u^{n+1}-u^n}{t_{n+1}-t_n},$$

in the Forward Euler, Backward Euler, and Crank–Nicolson schemes. Then we observe that the difference between the methods concerns which point this fraction approximates the derivative. Or in other words, at which point we sample the ODE. So far this has been the end points or the midpoint of $[t_n, t_{n+1}]$. However, we may choose any point $\tilde{t} \in [t_n, t_{n+1}]$. The difficulty is that evaluating the right-hand side $-au$ at an arbitrary point faces the same problem as in Sect. 1.1.4: the point value must be expressed by the discrete u quantities that we compute by the scheme, i.e., u^n and u^{n+1}. Following the averaging idea from Sect. 1.1.4, the value of u at an arbitrary point $\tilde{t}$ can be calculated as a *weighted average*, which generalizes the arithmetic mean $\frac{1}{2}u^n + \frac{1}{2}u^{n+1}$. The weighted average reads

$$u(\tilde{t}) \approx \theta u^{n+1} + (1-\theta)u^n, \tag{1.19}$$

where $\theta \in [0, 1]$ is a weighting factor. We can also express $\tilde{t}$ as a similar weighted average

$$\tilde{t} \approx \theta t_{n+1} + (1-\theta)t_n\,. \tag{1.20}$$

Let now the ODE hold at the point $\tilde{t} \in [t_n, t_{n+1}]$, approximate u' by the fraction $(u^{n+1}-u^n)/(t_{n+1}-t_n)$, and approximate the right-hand side $-au$ by the weighted average (1.19). The result is (1.17).

1.1.6 Constant Time Step

All schemes up to now have been formulated for a general non-uniform mesh in time: $t_0 < t_1 < \cdots < t_{N_t}$. Non-uniform meshes are highly relevant since one can use many points in regions where u varies rapidly, and fewer points in regions where u is slowly varying. This idea saves the total number of points and therefore makes it faster to compute the mesh function u^n. Non-uniform meshes are used together with *adaptive* methods that are able to adjust the time mesh during the computations (Sect. 3.2.11 applies adaptive methods).

However, a uniformly distributed set of mesh points is not only convenient, but also sufficient for many applications. Therefore, it is a very common choice. We shall present the finite difference schemes for a uniform point distribution $t_n = n\Delta t$, where Δt is the constant spacing between the mesh points, also referred to as the *time step*. The resulting formulas look simpler and are more well known.

Summary of schemes for constant time step

$$u^{n+1} = (1 - a\Delta t)u^n \quad \text{Forward Euler} \tag{1.21}$$

$$u^{n+1} = \frac{1}{1 + a\Delta t}u^n \quad \text{Backward Euler} \tag{1.22}$$

$$u^{n+1} = \frac{1 - \frac{1}{2}a\Delta t}{1 + \frac{1}{2}a\Delta t}u^n \quad \text{Crank–Nicolson} \tag{1.23}$$

$$u^{n+1} = \frac{1 - (1 - \theta)a\Delta t}{1 + \theta a\Delta t}u^n \quad \text{The } \theta\text{-rule} \tag{1.24}$$

It is not accidental that we focus on presenting the Forward Euler, Backward Euler, and Crank–Nicolson schemes. They complement each other with their different pros and cons, thus providing a useful collection of solution methods for many differential equation problems. The unifying notation of the θ-rule makes it convenient to work with all three methods through just one formula. This is particularly advantageous in computer implementations since one avoids if-else tests with formulas that have repetitive elements.

Test your understanding!

To check that key concepts are really understood, the reader is encouraged to apply the explained finite difference techniques to a slightly different equation. For this purpose, we recommend you do Exercise 4.2 now!

1.1.7 Mathematical Derivation of Finite Difference Formulas

The finite difference formulas for approximating the first derivative of a function have so far been somewhat justified through graphical illustrations in Figs. 1.3, 1.4, and 1.5. The task is to approximate the derivative at a point of a curve using only two function values. By drawing a straight line through the points, we have some approximation to the tangent of the curve and use the slope of this line as an approximation to the derivative. The slope can be computed by inspecting the figures.

However, we can alternatively derive the finite difference formulas by pure mathematics. The key tool for this approach is Taylor series, or more precisely, approximation of functions by lower-order Taylor polynomials. Given a function $f(x)$ that is sufficiently smooth (i.e., $f(x)$ has "enough derivatives"), a Taylor polynomial of degree m can be used to approximate the value of the function $f(x)$ if we know the values of f and its first m derivatives at some other point $x = a$. The formula for the Taylor polynomial reads

$$
\begin{aligned}
f(x) &\approx f(a) + f'(a)(x-a) + \frac{1}{2}f''(a)(x-a)^2 + \frac{1}{6}f'''(a)(x-a)^3 + \cdots \\
&+ \frac{1}{m!}\frac{df^{(m)}}{dx^m}(a)(x-a)^m .
\end{aligned} \tag{1.25}
$$

For a function of time, $f(t)$, related to a mesh with spacing Δt, we often need the Taylor polynomial approximation at $f(t_n \pm \Delta t)$ given f and its derivatives at $t = t_n$. Replacing x by $t_n + \Delta t$ and a by t_n gives

$$
\begin{aligned}
f(t_n + \Delta t) &\approx f(t_n) + f'(t_n)\Delta t + \frac{1}{2}f''(t_n)\Delta t^2 + \frac{1}{6}f'''(t_n)\Delta t^3 + \cdots \\
&+ \frac{1}{m!}\frac{df^{(m)}}{dx^m}(t_n)\Delta t^m .
\end{aligned} \tag{1.26}
$$

The forward difference We can use (1.26) to find an approximation for $f'(t_n)$ simply by solving with respect to this quantity:

$$
\begin{aligned}
f'(t_n) &\approx \frac{f(t_n + \Delta t) - f(t_n)}{\Delta t} - \frac{1}{2}f''(t_n)\Delta t - \frac{1}{6}f'''(t_n)\Delta t^2 + \cdots \\
&- \frac{1}{m!}\frac{df^{(m)}}{dx^m}(t_n)\Delta t^{m-1} .
\end{aligned} \tag{1.27}
$$

By letting $m \to \infty$, this formula is exact, but that is not so much of practical value. A more interesting observation is that all the power terms in Δt vanish as $\Delta t \to 0$, i.e., the formula

$$
f'(t_n) \approx \frac{f(t_n + \Delta t) - f(t_n)}{\Delta t} \tag{1.28}
$$

is exact in the limit $\Delta t \to 0$.

The interesting feature of (1.27) is that we have a measure of the error in the formula (1.28): the error is given by the extra terms on the right-hand side of (1.27). We assume that Δt is a small quantity ($\Delta t \ll 1$). Then $\Delta t^2 \ll \Delta t$, $\Delta t^3 \ll \Delta t^2$, and so on, which means that the first term is the dominating term. This first term reads $-\frac{1}{2}f''(t_n)\Delta t$ and can be taken as a measure of the error in the Forward Euler formula.

The backward difference To derive the backward difference, we use the Taylor polynomial approximation at $f(t_n - \Delta t)$:

$$
\begin{aligned}
f(t_n - \Delta t) \approx f(t_n) - f'(t_n)\Delta t + \frac{1}{2} f''(t_n)\Delta t^2 - \frac{1}{6} f'''(t_n)\Delta t^3 + \cdots \\
+ \frac{1}{m!}\frac{df^{(m)}}{dx^m}(t_n)\Delta t^m .
\end{aligned}
\tag{1.29}
$$

Solving with respect to $f'(t_n)$ gives

$$
\begin{aligned}
f'(t_n) \approx \frac{f(t_n) - f(t_n - \Delta t)}{\Delta t} + \frac{1}{2} f''(t_n)\Delta t - \frac{1}{6} f'''(t_n)\Delta t^2 + \cdots \\
- \frac{1}{m!}\frac{df^{(m)}}{dx^m}(t_n)\Delta t^{m-1} .
\end{aligned}
\tag{1.30}
$$

The term $\frac{1}{2} f''(t_n)\Delta t$ can be taken as a simple measure of the approximation error since it will dominate over the other terms as $\Delta t \to 0$.

The centered difference The centered difference approximates the derivative at $t_n + \frac{1}{2}\Delta t$. Let us write up the Taylor polynomial approximations to $f(t_n)$ and $f(t_{n+1})$ around $t_n + \frac{1}{2}\Delta t$:

$$
\begin{aligned}
f(t_n) \approx f\left(t_n + \frac{1}{2}\Delta t\right) - f'\left(t_n + \frac{1}{2}\Delta t\right)\frac{1}{2}\Delta t + f''\left(t_n + \frac{1}{2}\Delta t\right)\left(\frac{1}{2}\Delta t\right)^2 - \\
f'''\left(t_n + \frac{1}{2}\Delta t\right)\left(\frac{1}{2}\Delta t\right)^3 + \cdots
\end{aligned}
\tag{1.31}
$$

$$
\begin{aligned}
f(t_{n+1}) \approx f\left(t_n + \frac{1}{2}\Delta t\right) + f'\left(t_n + \frac{1}{2}\Delta t\right)\frac{1}{2}\Delta t + f''\left(t_n + \frac{1}{2}\Delta t\right)\left(\frac{1}{2}\Delta t\right)^2 + \\
f'''\left(t_n + \frac{1}{2}\Delta t\right)\left(\frac{1}{2}\Delta t\right)^3 + \cdots
\end{aligned}
\tag{1.32}
$$

Subtracting the first from the second gives

$$
f(t_{n+1}) - f(t_n) = f'\left(t_n + \frac{1}{2}\Delta t\right)\Delta t + 2f'''\left(t_n + \frac{1}{2}\Delta t\right)\left(\frac{1}{2}\Delta t\right)^3 + \cdots \tag{1.33}
$$

Solving with respect to $f'(t_n + \frac{1}{2}\Delta t)$ results in

$$
f'\left(t_n + \frac{1}{2}\Delta t\right) \approx \frac{f(t_{n+1}) - f(t_n)}{\Delta t} - \frac{1}{4} f'''\left(t_n + \frac{1}{2}\Delta t\right)\Delta t^2 + c \cdots \tag{1.34}
$$

This time the error measure goes like $\frac{1}{4} f''' \Delta t^2$, i.e., it is proportional to Δt^2 and not only Δt, which means that the error goes faster to zero as Δt is reduced. This means that the centered difference formula

$$
f'\left(t_n + \frac{1}{2}\Delta t\right) \approx \frac{f(t_{n+1}) - f(t_n)}{\Delta t} \tag{1.35}
$$

is more accurate than the forward and backward differences for small Δt.

1.1.8 Compact Operator Notation for Finite Differences

Finite difference formulas can be tedious to write and read, especially for differential equations with many terms and many derivatives. To save space and help the reader spot the nature of the difference approximations, we introduce a compact notation. For a function $u(t)$, a forward difference approximation is denoted by the D_t^+ operator and written as

$$[D_t^+ u]^n = \frac{u^{n+1} - u^n}{\Delta t} \quad \left(\approx \frac{d}{dt} u(t_n)\right). \tag{1.36}$$

The notation consists of an operator that approximates differentiation with respect to an independent variable, here t. The operator is built of the symbol D, with the independent variable as subscript and a superscript denoting the type of difference. The superscript $^+$ indicates a forward difference. We place square brackets around the operator and the function it operates on and specify the mesh point, where the operator is acting, by a superscript after the closing bracket.

The corresponding operator notation for a centered difference and a backward difference reads

$$[D_t u]^n = \frac{u^{n+\frac{1}{2}} - u^{n-\frac{1}{2}}}{\Delta t} \approx \frac{d}{dt} u(t_n), \tag{1.37}$$

and

$$[D_t^- u]^n = \frac{u^n - u^{n-1}}{\Delta t} \approx \frac{d}{dt} u(t_n). \tag{1.38}$$

Note that the superscript $^-$ denotes the backward difference, while no superscript implies a central difference.

An averaging operator is also convenient to have:

$$[\overline{u}^t]^n = \frac{1}{2}(u^{n-\frac{1}{2}} + u^{n+\frac{1}{2}}) \approx u(t_n) \tag{1.39}$$

The superscript t indicates that the average is taken along the time coordinate. The common average $(u^n + u^{n+1})/2$ can now be expressed as $[\overline{u}^t]^{n+\frac{1}{2}}$. (When also spatial coordinates enter the problem, we need the explicit specification of the coordinate after the bar.)

With our compact notation, the Backward Euler finite difference approximation to $u' = -au$ can be written as

$$[D_t^- u]^n = -au^n.$$

In difference equations we often place the square brackets around the whole equation, to indicate at which mesh point the equation applies, since each term must be approximated at the same point:

$$[D_t^- u = -au]^n. \tag{1.40}$$

Similarly, the Forward Euler scheme takes the form

$$[D_t^+ u = -au]^n, \tag{1.41}$$

while the Crank–Nicolson scheme is written as

$$[D_t u = -a\overline{u}^t]^{n+\frac{1}{2}}. \tag{1.42}$$

Question

By use of (1.37) and (1.39), are you able to write out the expressions in (1.42) to verify that it is indeed the Crank–Nicolson scheme?

The θ-rule can be specified in operator notation by

$$[\bar{D}_t u = -a\overline{u}^{t,\theta}]^{n+\theta}. \tag{1.43}$$

We define a new time difference

$$[\bar{D}_t u]^{n+\theta} = \frac{u^{n+1} - u^n}{t^{n+1} - t^n}, \tag{1.44}$$

to be applied at the time point $t_{n+\theta} \approx \theta t_n + (1-\theta)t_{n+1}$. This weighted average gives rise to the *weighted averaging operator*

$$[\overline{u}^{t,\theta}]^{n+\theta} = (1-\theta)u^n + \theta u^{n+1} \approx u(t_{n+\theta}), \tag{1.45}$$

where $\theta \in [0,1]$ as usual. Note that for $\theta = \frac{1}{2}$ we recover the standard centered difference and the standard arithmetic mean. The idea in (1.43) is to sample the equation at $t_{n+\theta}$, use a non-symmetric difference at that point $[\bar{D}_t u]^{n+\theta}$, and a weighted (non-symmetric) mean value.

An alternative and perhaps clearer notation is

$$[D_t u]^{n+\frac{1}{2}} = \theta[-au]^{n+1} + (1-\theta)[-au]^n.$$

Looking at the various examples above and comparing them with the underlying differential equations, we see immediately which difference approximations that have been used and at which point they apply. Therefore, the compact notation effectively communicates the reasoning behind turning a differential equation into a difference equation.

1.2 Implementations

We want to make a computer program for solving

$$u'(t) = -au(t), \quad t \in (0,T], \quad u(0) = I,$$

by finite difference methods. The program should also display the numerical solution as a curve on the screen, preferably together with the exact solution.

All programs referred to in this section are found in the `src/alg`[1] directory (we use the classical Unix term *directory* for what many others nowadays call *folder*).

[1] http://tinyurl.com/ofkw6kc/alg

Mathematical problem We want to explore the Forward Euler scheme, the Backward Euler, and the Crank–Nicolson schemes applied to our model problem. From an implementational point of view, it is advantageous to implement the θ-rule

$$u^{n+1} = \frac{1 - (1-\theta)a\Delta t}{1 + \theta a\Delta t}u^n,$$

since it can generate the three other schemes by various choices of θ: $\theta = 0$ for Forward Euler, $\theta = 1$ for Backward Euler, and $\theta = 1/2$ for Crank–Nicolson. Given a, $u^0 = I$, T, and Δt, our task is to use the θ-rule to compute $u^1, u^2, \ldots, u^{N_t}$, where $t_{N_t} = N_t\Delta t$, and N_t the closest integer to $T/\Delta t$.

1.2.1 Computer Language: Python

Any programming language can be used to generate the u^{n+1} values from the formula above. However, in this document we shall mainly make use of Python. There are several good reasons for this choice:

- Python has a very clean, readable syntax (often known as "executable pseudo-code").
- Python code is very similar to MATLAB code (and MATLAB has a particularly widespread use for scientific computing).
- Python is a full-fledged, very powerful programming language.
- Python is similar to C++, but is much simpler to work with and results in more reliable code.
- Python has a rich set of modules for scientific computing, and its popularity in scientific computing is rapidly growing.
- Python was made for being combined with compiled languages (C, C++, Fortran), so that existing numerical software can be reused, and thereby easing high computational performance with new implementations.
- Python has extensive support for administrative tasks needed when doing large-scale computational investigations.
- Python has extensive support for graphics (visualization, user interfaces, web applications).

Learning Python is easy. Many newcomers to the language will probably learn enough from the forthcoming examples to perform their own computer experiments. The examples start with simple Python code and gradually make use of more powerful constructs as we proceed. Unless it is inconvenient for the problem at hand, our Python code is made as close as possible to MATLAB code for easy transition between the two languages.

The coming programming examples assumes familiarity with variables, for loops, lists, arrays, functions, positional arguments, and keyword (named) arguments. A background in basic MATLAB programming is often enough to understand Python examples. Readers who feel the Python examples are too hard to follow will benefit from reading a tutorial, e.g.,

- The Official Python Tutorial[2]
- Python Tutorial on tutorialspoint.com[3]
- Interactive Python tutorial site[4]
- A Beginner's Python Tutorial[5]

The author also has a comprehensive book [8] that teaches scientific programming with Python from the ground up.

1.2.2 Making a Solver Function

We choose to have an array u for storing the u^n values, $n = 0, 1, \ldots, N_t$. The algorithmic steps are

1. initialize u^0
2. for $t = t_n$, $n = 1, 2, \ldots, N_t$: compute u_n using the θ-rule formula

An implementation of a numerical algorithm is often referred to as a *solver*. We shall now make a solver for our model problem and realize the solver as a Python function. The function must take the input data I, a, T, Δt, and θ of the problem as arguments and return the solution as arrays u and t for u^n and t^n, $n = 0, \ldots, N_t$. The solver function used as

```
u, t = solver(I, a, T, dt, theta)
```

One can now easily plot u versus t to visualize the solution.

The function solver may look as follows in Python:

```
from numpy import *

def solver(I, a, T, dt, theta):
    """Solve u'=-a*u, u(0)=I, for t in (0,T] with steps of dt."""
    Nt = int(T/dt)              # no of time intervals
    T = Nt*dt                   # adjust T to fit time step dt
    u = zeros(Nt+1)             # array of u[n] values
    t = linspace(0, T, Nt+1)    # time mesh

    u[0] = I                    # assign initial condition
    for n in range(0, Nt):      # n=0,1,...,Nt-1
        u[n+1] = (1 - (1-theta)*a*dt)/(1 + theta*dt*a)*u[n]
    return u, t
```

The numpy library contains a lot of functions for array computing. Most of the function names are similar to what is found in the alternative scientific computing language MATLAB. Here we make use of

- zeros(Nt+1) for creating an array of size Nt+1 and initializing the elements to zero

[2] http://docs.python.org/2/tutorial/
[3] http://www.tutorialspoint.com/python/
[4] http://www.learnpython.org/
[5] http://en.wikibooks.org/wiki/A_Beginner's_Python_Tutorial

- `linspace(0, T, Nt+1)` for creating an array with `Nt+1` coordinates uniformly distributed between 0 and T

The `for` loop deserves a comment, especially for newcomers to Python. The construction `range(0, Nt, s)` generates all integers from 0 to `Nt` in steps of `s`, *but not including* `Nt`. Omitting `s` means `s=1`. For example, `range(0, 6, 3)` gives 0 and 3, while `range(0, 6)` generates the list `[0, 1, 2, 3, 4, 5]`. Our loop implies the following assignments to `u[n+1]`: `u[1]`, `u[2]`, ..., `u[Nt]`, which is what we want since `u` has length `Nt+1`. The first index in Python arrays or lists is *always* 0 and the last is then `len(u)-1` (the length of an array `u` is obtained by `len(u)` or `u.size`).

1.2.3 Integer Division

The shown implementation of the `solver` may face problems and wrong results if `T`, `a`, `dt`, and `theta` are given as integers (see Exercises 1.3 and 1.4). The problem is related to *integer division* in Python (as in Fortran, C, C++, and many other computer languages!): `1/2` becomes `0`, while `1.0/2`, `1/2.0`, or `1.0/2.0` all become `0.5`. So, it is enough that at least the nominator or the denominator is a real number (i.e., a `float` object) to ensure a correct mathematical division. Inserting a conversion `dt = float(dt)` guarantees that `dt` is `float`.

Another problem with computing $N_t = T/\Delta t$ is that we should round N_t to the nearest integer. With `Nt = int(T/dt)` the `int` operation picks the largest integer smaller than `T/dt`. Correct mathematical rounding as known from school is obtained by

```
Nt = int(round(T/dt))
```

The complete version of our improved, safer `solver` function then becomes

```
from numpy import *

def solver(I, a, T, dt, theta):
    """Solve u'=-a*u, u(0)=I, for t in (0,T] with steps of dt."""
    dt = float(dt)               # avoid integer division
    Nt = int(round(T/dt))        # no of time intervals
    T = Nt*dt                    # adjust T to fit time step dt
    u = zeros(Nt+1)              # array of u[n] values
    t = linspace(0, T, Nt+1)     # time mesh

    u[0] = I                     # assign initial condition
    for n in range(0, Nt):       # n=0,1,...,Nt-1
        u[n+1] = (1 - (1-theta)*a*dt)/(1 + theta*dt*a)*u[n]
    return u, t
```

1.2.4 Doc Strings

Right below the header line in the `solver` function there is a Python string enclosed in triple double quotes `"""`. The purpose of this string object is to document what the function does and what the arguments are. In this case the necessary documentation does not span more than one line, but with triple double quoted strings the text may span several lines:

```
def solver(I, a, T, dt, theta):
    """
    Solve

        u'(t) = -a*u(t),

    with initial condition u(0)=I, for t in the time interval
    (0,T]. The time interval is divided into time steps of
    length dt.

    theta=1 corresponds to the Backward Euler scheme, theta=0
    to the Forward Euler scheme, and theta=0.5 to the Crank-
    Nicolson method.
    """
    ...
```

Such documentation strings appearing right after the header of a function are called *doc strings*. There are tools that can automatically produce nicely formatted documentation by extracting the definition of functions and the contents of doc strings.

It is strongly recommended to equip any function with a doc string, unless the purpose of the function is not obvious. Nevertheless, the forthcoming text deviates from this rule if the function is explained in the text.

1.2.5 Formatting Numbers

Having computed the discrete solution `u`, it is natural to look at the numbers:

```
# Write out a table of t and u values:
for i in range(len(t)):
    print t[i], u[i]
```

This compact `print` statement unfortunately gives less readable output because the `t` and `u` values are not aligned in nicely formatted columns. To fix this problem, we recommend to use the *printf format*, supported in most programming languages inherited from C. Another choice is Python's recent *format string syntax*. Both kinds of syntax are illustrated below.

Writing `t[i]` and `u[i]` in two nicely formatted columns is done like this with the printf format:

```
print 't=%6.3f u=%g' % (t[i], u[i])
```

The percentage signs signify "slots" in the text where the variables listed at the end of the statement are inserted. For each "slot" one must specify a format for how the variable is going to appear in the string: `f` for float (with 6 decimals), `s` for pure text, `d` for an integer, `g` for a real number written as compactly as possible, `9.3E` for scientific notation with three decimals in a field of width 9 characters (e.g., `-1.351E-2`), or `.2f` for standard decimal notation with two decimals formatted with minimum width. The printf syntax provides a quick way of formatting tabular output of numbers with full control of the layout.

The alternative *format string syntax* looks like

```
print 't={t:6.3f} u={u:g}'.format(t=t[i], u=u[i])
```

As seen, this format allows logical names in the "slots" where `t[i]` and `u[i]` are to be inserted. The "slots" are surrounded by curly braces, and the logical name is followed by a colon and then the printf-like specification of how to format real numbers, integers, or strings.

1.2.6 Running the Program

The function and main program shown above must be placed in a file, say with name `decay_v1.py`[6] (`v1` for 1st version of this program). Make sure you write the code with a suitable text editor (Gedit, Emacs, Vim, Notepad++, or similar). The program is run by executing the file this way:

```
Terminal

Terminal> python decay_v1.py
```

The text `Terminal>` just indicates a prompt in a Unix/Linux or DOS terminal window. After this prompt, which may look different in your terminal window (depending on the terminal application and how it is set up), commands like `python decay_v1.py` can be issued. These commands are interpreted by the operating system.

We strongly recommend to run Python programs within the IPython shell. First start IPython by typing `ipython` in the terminal window. Inside the IPython shell, our program `decay_v1.py` is run by the command `run decay_v1.py`:

```
Terminal

Terminal> ipython

In [1]: run decay_v1.py
t= 0.000 u=1
t= 0.800 u=0.384615
t= 1.600 u=0.147929
t= 2.400 u=0.0568958
t= 3.200 u=0.021883
t= 4.000 u=0.00841653
t= 4.800 u=0.00323713
t= 5.600 u=0.00124505
t= 6.400 u=0.000478865
t= 7.200 u=0.000184179
t= 8.000 u=7.0838e-05
```

[6] http://tinyurl.com/ofkw6kc/alg/decay_v1.py

The advantage of running programs in IPython are many, but here we explicitly mention a few of the most useful features:

- previous commands are easily recalled with the up arrow,
- `%pdb` turns on a debugger so that variables can be examined if the program aborts (due to a Python exception),
- output of commands are stored in variables,
- the computing time spent on a set of statements can be measured with the `%timeit` command,
- any operating system command can be executed,
- modules can be loaded automatically and other customizations can be performed when starting IPython

Although running programs in IPython is strongly recommended, most execution examples in the forthcoming text use the standard Python shell with prompt `>>>` and run programs through a typesetting like

```
Terminal> python programname
```

The reason is that such typesetting makes the text more compact in the vertical direction than showing sessions with IPython syntax.

1.2.7 Plotting the Solution

Having the `t` and `u` arrays, the approximate solution `u` is visualized by the intuitive command `plot(t, u)`:

```
from matplotlib.pyplot import *
plot(t, u)
show()
```

It will be illustrative to also plot the exact solution $u_{\mathrm{e}}(t) = Ie^{-at}$ for comparison. We first need to make a Python function for computing the exact solution:

```
def u_exact(t, I, a):
    return I*exp(-a*t)
```

It is tempting to just do

```
u_e = u_exact(t, I, a)
plot(t, u, t, u_e)
```

However, this is not exactly what we want: the `plot` function draws straight lines between the discrete points `(t[n], u_e[n])` while $u_{\mathrm{e}}(t)$ varies as an exponential function between the mesh points. The technique for showing the "exact" variation of $u_e(t)$ between the mesh points is to introduce a very fine mesh for $u_e(t)$:

```
t_e = linspace(0, T, 1001)      # fine mesh
u_e = u_exact(t_e, I, a)
```

We can also plot the curves with different colors and styles, e.g.,

```
plot(t_e, u_e, 'b-',           # blue line for u_e
     t,   u,   'r--o')         # red dashes w/circles
```

With more than one curve in the plot we need to associate each curve with a legend. We also want appropriate names on the axes, a title, and a file containing the plot as an image for inclusion in reports. The Matplotlib package (`matplotlib.pyplot`) contains functions for this purpose. The names of the functions are similar to the plotting functions known from MATLAB. A complete function for creating the comparison plot becomes

```
from matplotlib.pyplot import *

def plot_numerical_and_exact(theta, I, a, T, dt):
    """Compare the numerical and exact solution in a plot."""
    u, t = solver(I=I, a=a, T=T, dt=dt, theta=theta)

    t_e = linspace(0, T, 1001)        # fine mesh for u_e
    u_e = u_exact(t_e, I, a)

    plot(t,   u,   'r--o',            # red dashes w/circles
         t_e, u_e, 'b-')              # blue line for exact sol.
    legend(['numerical', 'exact'])
    xlabel('t')
    ylabel('u')
    title('theta=%g, dt=%g' % (theta, dt))
    savefig('plot_%s_%g.png' % (theta, dt))

plot_numerical_and_exact(I=1, a=2, T=8, dt=0.8, theta=1)
show()
```

Note that `savefig` here creates a PNG file whose name includes the values of θ and Δt so that we can easily distinguish files from different runs with θ and Δt.

The complete code is found in the file `decay_v2.py`[7]. The resulting plot is shown in Fig. 1.6. As seen, there is quite some discrepancy between the exact and the numerical solution. Fortunately, the numerical solution approaches the exact one as Δt is reduced.

1.2.8 Verifying the Implementation

It is easy to make mistakes while deriving and implementing numerical algorithms, so we should never believe in the solution before it has been thoroughly verified.

[7] http://tinyurl.com/ofkw6kc/alg/decay_v2.py

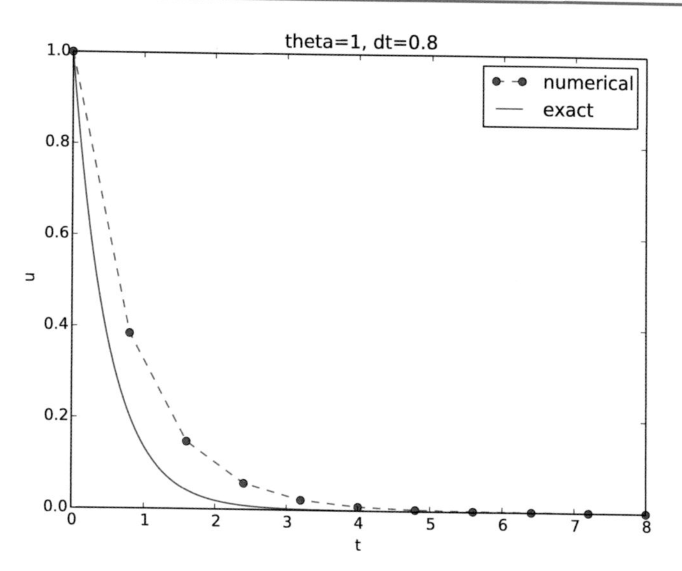

Fig. 1.6 Comparison of numerical and exact solution

Verification and validation

The purpose of *verifying* a program is to bring evidence for the property that there are no errors in the implementation. A related term, *validate* (and *validation*), addresses the question if the ODE model is a good representation of the phenomena we want to simulate. To remember the difference between verification and validation, verification is about *solving the equations right*, while validation is about *solving the right equations*. We must always perform a verification before it is meaningful to believe in the computations and perform validation (which compares the program results with physical experiments or observations).

The most obvious idea for verification in our case is to compare the numerical solution with the exact solution, when that exists. This is, however, not a particularly good method. The reason is that there will always be a discrepancy between these two solutions, due to numerical approximations, and we cannot precisely quantify the approximation errors. The open question is therefore whether we have the mathematically correct discrepancy or if we have another, maybe small, discrepancy due to both an approximation error *and* an error in the implementation. It is thus impossible to judge whether the program is correct or not by just looking at the graphs in Fig. 1.6.

To avoid mixing the unavoidable numerical approximation errors and the undesired implementation errors, we should try to make tests where we have some exact computation of the discrete solution or at least parts of it. Examples will show how this can be done.

Running a few algorithmic steps by hand The simplest approach to produce a correct non-trivial reference solution for the discrete solution u, is to compute a few steps of the algorithm by hand. Then we can compare the hand calculations with numbers produced by the program.

A straightforward approach is to use a calculator and compute u^1, u^2, and u^3. With $I = 0.1$, $\theta = 0.8$, and $\Delta t = 0.8$ we get

$$A \equiv \frac{1-(1-\theta)a\Delta t}{1+\theta a\Delta t} = 0.298245614035$$

$$u^1 = AI = 0.0298245614035,$$
$$u^2 = Au^1 = 0.00889504462912,$$
$$u^3 = Au^2 = 0.00265290804728$$

Comparison of these manual calculations with the result of the `solver` function is carried out in the function

```
def test_solver_three_steps():
    """Compare three steps with known manual computations."""
    theta = 0.8; a = 2; I = 0.1; dt = 0.8
    u_by_hand = array([I,
                       0.0298245614035,
                       0.00889504462912,
                       0.00265290804728])

    Nt = 3  # number of time steps
    u, t = solver(I=I, a=a, T=Nt*dt, dt=dt, theta=theta)

    tol = 1E-15  # tolerance for comparing floats
    diff = abs(u - u_by_hand).max()
    success = diff < tol
    assert success
```

The `test_solver_three_steps` function follows widely used conventions for *unit testing*. By following such conventions we can at a later stage easily execute a big test suite for our software. That is, after a small modification is made to the program, we can by typing just a short command, run through a large number of tests to check that the modifications do not break any computations. The conventions boil down to three rules:

- The test function name must start with `test_` and the function cannot take any arguments.
- The test must end up in a boolean expression that is `True` if the test was passed and `False` if it failed.
- The function must run `assert` on the boolean expression, resulting in program abortion (due to an `AssertionError` exception) if the test failed.

A typical `assert` statement is to check that a computed result `c` equals the expected value `e`: `assert c == e`. However, since real numbers are stored in a computer using only 64 units, most numbers will feature a small rounding error, typically of size 10^{-16}. That is, real numbers on a computer have finite precision. When doing arithmetics with finite precision numbers, the rounding errors may accumulate or not, depending on the algorithm. It does not make sense to test `c == e`, since a small rounding error will cause the test to fail. Instead, we use an equality with *tolerance* `tol`: `abs(e - c) < tol`. The `test_solver_three_steps` functions applies this type of test with a tolerance 01^{-15}.

The main program can routinely run the verification test prior to solving the real problem:

```
test_solver_three_steps()
plot_numerical_and_exact(I=1, a=2, T=8, dt=0.8, theta=1)
show()
```

(Rather than calling `test_*()` functions explicitly, one will normally ask a testing framework like nose or pytest to find and run such functions.) The complete program including the verification above is found in the file `decay_v3.py`[8].

1.2.9 Computing the Numerical Error as a Mesh Function

Now that we have some evidence for a correct implementation, we are in position to compare the computed u^n values in the `u` array with the exact u values at the mesh points, in order to study the error in the numerical solution.

A natural way to compare the exact and discrete solutions is to calculate their difference as a mesh function for the error:

$$e^n = u_e(t_n) - u^n, \quad n = 0, 1, \ldots, N_t\,. \tag{1.46}$$

We may view the mesh function $u_e^n = u_e(t_n)$ as a representation of the continuous function $u_e(t)$ defined for all $t \in [0, T]$. In fact, u_e^n is often called the *representative* of u_e on the mesh. Then, $e^n = u_e^n - u^n$ is clearly the difference of two mesh functions.

The error mesh function e^n can be computed by

```
u, t = solver(I, a, T, dt, theta)  # Numerical sol.
u_e = u_exact(t, I, a)             # Representative of exact sol.
e = u_e - u
```

Note that the mesh functions `u` and `u_e` are represented by arrays and associated with the points in the array `t`.

Array arithmetics

The last statements

```
u_e = u_exact(t, I, a)
e = u_e - u
```

demonstrate some standard examples of array arithmetics: `t` is an array of mesh points that we pass to `u_exact`. This function evaluates `-a*t`, which is a scalar times an array, meaning that the scalar is multiplied with each array element. The result is an array, let us call it `tmp1`. Then `exp(tmp1)` means applying the exponential function to each element in `tmp1`, giving an array, say `tmp2`.

[8] http://tinyurl.com/ofkw6kc/alg/decay_v3.py

Finally, `I*tmp2` is computed (scalar times array) and `u_e` refers to this array returned from `u_exact`. The expression `u_e - u` is the difference between two arrays, resulting in a new array referred to by `e`.

Replacement of array element computations inside a loop by array arithmetics is known as *vectorization*.

1.2.10 Computing the Norm of the Error Mesh Function

Instead of working with the error e^n on the entire mesh, we often want a single number expressing the size of the error. This is obtained by taking the norm of the error function.

Let us first define norms of a function $f(t)$ defined for all $t \in [0, T]$. Three common norms are

$$||f||_{L^2} = \left(\int_0^T f(t)^2 dt\right)^{1/2}, \tag{1.47}$$

$$||f||_{L^1} = \int_0^T |f(t)|dt, \tag{1.48}$$

$$||f||_{L^\infty} = \max_{t\in[0,T]} |f(t)|. \tag{1.49}$$

The L^2 norm (1.47) ("L-two norm") has nice mathematical properties and is the most popular norm. It is a generalization of the well-known Eucledian norm of vectors to functions. The L^1 norm looks simpler and more intuitive, but has less nice mathematical properties compared to the two other norms, so it is much less used in computations. The L^∞ is also called the max norm or the supremum norm and is widely used. It focuses on a single point with the largest value of $|f|$, while the other norms measure average behavior of the function.

In fact, there is a whole family of norms,

$$||f||_{L^p} = \left(\int_0^T f(t)^p dt\right)^{1/p}, \tag{1.50}$$

with p real. In particular, $p = 1$ corresponds to the L^1 norm above while $p = \infty$ is the L^∞ norm.

Numerical computations involving mesh functions need corresponding norms. Given a set of function values, f^n, and some associated mesh points, t_n, a numerical integration rule can be used to calculate the L^2 and L^1 norms defined above. Imagining that the mesh function is extended to vary linearly between the mesh points, the Trapezoidal rule is in fact an exact integration rule. A possible modification of the L^2 norm for a mesh function f^n on a uniform mesh with spacing Δt

is therefore the well-known Trapezoidal integration formula

$$||f^n|| = \left(\Delta t\left(\frac{1}{2}(f^0)^2 + \frac{1}{2}(f^{N_t})^2 + \sum_{n=1}^{N_t-1}(f^n)^2\right)\right)^{1/2}$$

A common approximation of this expression, motivated by the convenience of having a simpler formula, is

$$||f^n||_{\ell^2} = \left(\Delta t\sum_{n=0}^{N_t}(f^n)^2\right)^{1/2}.$$

This is called the discrete L^2 norm and denoted by ℓ^2. If $||f||_{\ell^2}^2$ (i.e., the square of the norm) is used instead of the Trapezoidal integration formula, the error is $\Delta t((f^0)^2 + (f^{N_t})^2)/2$. This means that the weights at the end points of the mesh function are perturbed, but as $\Delta t \to 0$, the error from this perturbation goes to zero. As long as we are consistent and stick to one kind of integration rule for the norm of a mesh function, the details and accuracy of this rule is of no concern.

The three discrete norms for a mesh function f^n, corresponding to the L^2, L^1, and L^∞ norms of $f(t)$ defined above, are defined by

$$||f^n||_{\ell^2} = \left(\Delta t\sum_{n=0}^{N_t}(f^n)^2\right)^{1/2}, \tag{1.51}$$

$$||f^n||_{\ell^1} = \Delta t\sum_{n=0}^{N_t}|f^n|, \tag{1.52}$$

$$||f^n||_{\ell^\infty} = \max_{0\le n\le N_t}|f^n|. \tag{1.53}$$

Note that the L^2, L^1, ℓ^2, and ℓ^1 norms depend on the length of the interval of interest (think of $f = 1$, then the norms are proportional to $\sqrt{T}$ or T). In some applications it is convenient to think of a mesh function as just a vector of function values without any relation to the interval $[0, T]$. Then one can replace Δt by T/N_t and simply drop T (which is just a common scaling factor in the norm, independent of the vector of function values). Moreover, people prefer to divide by the total length of the vector, $N_t + 1$, instead of N_t. This reasoning gives rise to the *vector norms* for a vector $f = (f_0, \ldots, f_N)$:

$$||f||_2 = \left(\frac{1}{N+1}\sum_{n=0}^{N}(f_n)^2\right)^{1/2}, \tag{1.54}$$

$$||f||_1 = \frac{1}{N+1}\sum_{n=0}^{N}|f_n|, \tag{1.55}$$

$$||f||_{\ell^\infty} = \max_{0\le n\le N}|f_n|. \tag{1.56}$$

Here we have used the common vector component notation with subscripts (f_n) and N as length. We will mostly work with mesh functions and use the discrete ℓ^2 norm (1.51) or the max norm ℓ^∞ (1.53), but the corresponding vector norms (1.54)–(1.56) are also much used in numerical computations, so it is important to know the different norms and the relations between them.

A single number that expresses the size of the numerical error will be taken as $||e^n||_{\ell^2}$ and called E:

$$E = \sqrt{\Delta t \sum_{n=0}^{N_t} (e^n)^2} \tag{1.57}$$

The corresponding Python code, using array arithmetics, reads

```
E = sqrt(dt*sum(e**2))
```

The `sum` function comes from `numpy` and computes the sum of the elements of an array. Also the `sqrt` function is from `numpy` and computes the square root of each element in the array argument.

Scalar computing Instead of doing array computing `sqrt(dt*sum(e**2))` we can compute with one element at a time:

```
m = len(u)      # length of u array (alt: u.size)
u_e = zeros(m)
t = 0
for i in range(m):
    u_e[i] = u_exact(t, a, I)
    t = t + dt
e = zeros(m)
for i in range(m):
    e[i] = u_e[i] - u[i]
s = 0  # summation variable
for i in range(m):
    s = s + e[i]**2
error = sqrt(dt*s)
```

Such element-wise computing, often called *scalar* computing, takes more code, is less readable, and runs much slower than what we can achieve with array computing.

1.2.11 Experiments with Computing and Plotting

Let us write down a new function that wraps up the computation and all the plotting statements used for comparing the exact and numerical solutions. This function can be called with various θ and Δt values to see how the error depends on the method and mesh resolution.

```
def explore(I, a, T, dt, theta=0.5, makeplot=True):
    """
    Run a case with the solver, compute error measure,
    and plot the numerical and exact solutions (if makeplot=True).
    """
```

```
    u, t = solver(I, a, T, dt, theta)    # Numerical solution
    u_e = u_exact(t, I, a)
    e = u_e - u
    E = sqrt(dt*sum(e**2))
    if makeplot:
        figure()                         # create new plot
        t_e = linspace(0, T, 1001)       # fine mesh for u_e
        u_e = u_exact(t_e, I, a)
        plot(t,   u,   'r--o')           # red dashes w/circles
        plot(t_e, u_e, 'b-')             # blue line for exact sol.
        legend(['numerical', 'exact'])
        xlabel('t')
        ylabel('u')
        title('theta=%g, dt=%g' % (theta, dt))
        theta2name = {0: 'FE', 1: 'BE', 0.5: 'CN'}
        savefig('%s_%g.png' % (theta2name[theta], dt))
        savefig('%s_%g.pdf' % (theta2name[theta], dt))
        show()
    return E
```

The `figure()` call is key: without it, a new `plot` command will draw the new pair of curves in the same plot window, while we want the different pairs to appear in separate windows and files. Calling `figure()` ensures this.

Instead of including the θ value in the filename to implicitly inform about the applied method, the code utilizes a little Python dictionary that maps each relevant θ value to a corresponding acronym for the method name (FE, BE, or CN):

```
theta2name = {0: 'FE', 1: 'BE', 0.5: 'CN'}
savefig('%s_%g.png' % (theta2name[theta], dt))
```

The `explore` function stores the plot in two different image file formats: PNG and PDF. The PNG format is suitable for being included in HTML documents, while the PDF format provides higher quality for LaTeX (i.e., PDFLaTeX) documents. Frequently used viewers for these image files on Unix systems are `gv` (comes with Ghostscript) for the PDF format and `display` (from the ImageMagick software suite) for PNG files:

Terminal

```
Terminal> gv BE_0.5.pdf
Terminal> display BE_0.5.png
```

A main program may run a loop over the three methods (given by their corresponding θ values) and call `explore` to compute errors and make plots:

```
def main(I, a, T, dt_values, theta_values=(0, 0.5, 1)):
    print 'theta   dt        error'  # Column headings in table
    for theta in theta_values:
        for dt in dt_values:
            E = explore(I, a, T, dt, theta, makeplot=True)
            print '%4.1f %6.2f: %12.3E' % (theta, dt, E)

main(I=1, a=2, T=5, dt_values=[0.4, 0.04])
```

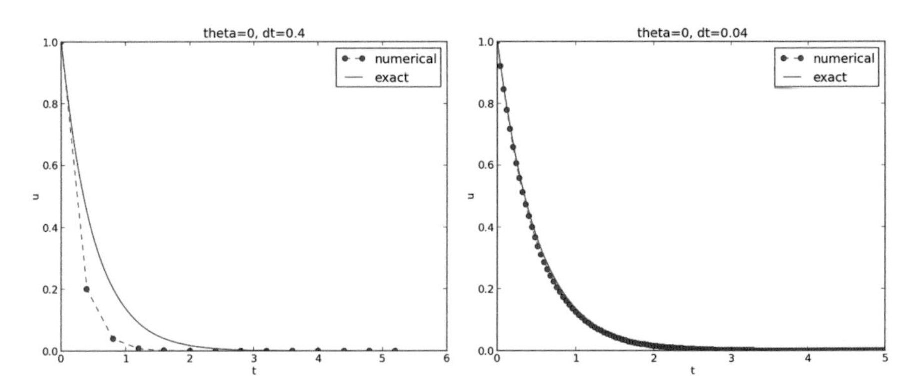

Fig. 1.7 The Forward Euler scheme for two values of the time step

The file `decay_plot_mpl.py`[9] contains the complete code with the functions above. Running this program results in

```
Terminal> python decay_plot_mpl.py
theta     dt       error
 0.0    0.40:     2.105E-01
 0.0    0.04:     1.449E-02
 0.5    0.40:     3.362E-02
 0.5    0.04:     1.887E-04
 1.0    0.40:     1.030E-01
 1.0    0.04:     1.382E-02
```

We observe that reducing Δt by a factor of 10 increases the accuracy for all three methods. We also see that the combination of $\theta = 0.5$ and a small time step $\Delta t = 0.04$ gives a much more accurate solution, and that $\theta = 0$ and $\theta = 1$ with $\Delta t = 0.4$ result in the least accurate solutions.

Figure 1.7 demonstrates that the numerical solution produced by the Forward Euler method with $\Delta t = 0.4$ clearly lies below the exact curve, but that the accuracy improves considerably by reducing the time step by a factor of 10.

The behavior of the two other schemes is shown in Figs. 1.8 and 1.9. Crank–Nicolson is obviously the most accurate scheme from this visual point of view.

Combining plot files Mounting two PNG files beside each other, as done in Figs. 1.7–1.9, is easily carried out by the `montage`[10] program from the ImageMagick suite:

```
Terminal> montage -background white -geometry 100% -tile 2x1 \
          FE_0.4.png FE_0.04.png FE1.png
Terminal> convert -trim FE1.png FE1.png
```

[9] http://tinyurl.com/ofkw6kc/alg/decay_plot_mpl.py
[10] http://www.imagemagick.org/script/montage.php

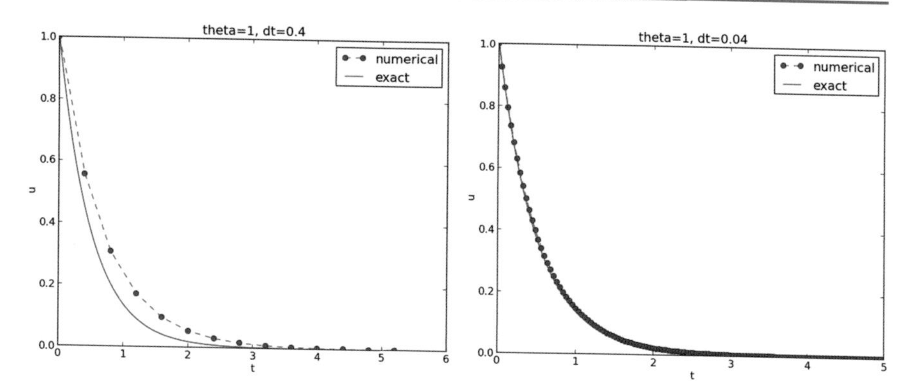

Fig. 1.8 The Backward Euler scheme for two values of the time step

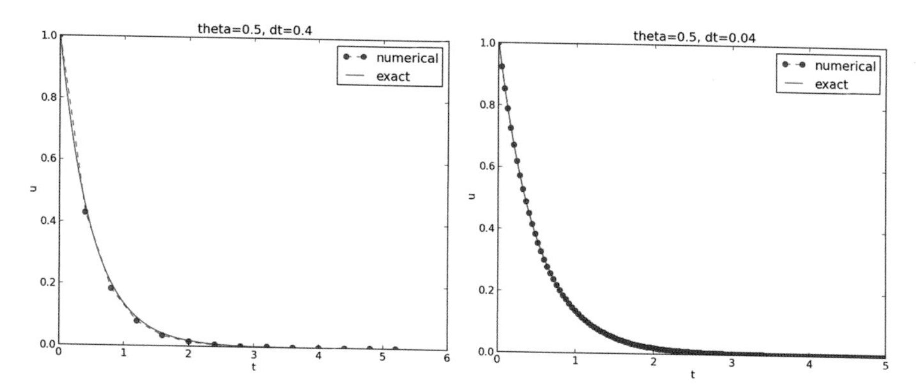

Fig. 1.9 The Crank–Nicolson scheme for two values of the time step

The `-geometry` argument is used to specify the size of the image. Here, we preserve the individual sizes of the images. The `-tile HxV` option specifies `H` images in the horizontal direction and `V` images in the vertical direction. A series of image files to be combined are then listed, with the name of the resulting combined image, here `FE1.png` at the end. The `convert -trim` command removes surrounding white areas in the figure (an operation usually known as *cropping* in image manipulation programs).

For LaTeX reports it is not recommended to use `montage` and PNG files as the result has too low resolution. Instead, plots should be made in the PDF format and combined using the `pdftk`, `pdfnup`, and `pdfcrop` tools (on Linux/Unix):

```
Terminal> pdftk FE_0.4.png FE_0.04.png output tmp.pdf
Terminal> pdfnup --nup 2x1 --outfile tmp.pdf tmp.pdf
Terminal> pdfcrop tmp.pdf FE1.png  # output in FE1.png
```

Here, `pdftk` combines images into a multi-page PDF file, `pdfnup` combines the images in individual pages to a table of images (pages), and `pdfcrop` removes white margins in the resulting combined image file.

Plotting with SciTools The SciTools package[11] provides a unified plotting interface, called Easyviz, to many different plotting packages, including Matplotlib, Gnuplot, Grace, MATLAB, VTK, OpenDX, and VisIt. The syntax is very similar to that of Matplotlib and MATLAB. In fact, the plotting commands shown above look the same in SciTool's Easyviz interface, apart from the import statement, which reads

```
from scitools.std import *
```

This statement performs a `from numpy import *` as well as an import of the most common pieces of the Easyviz (`scitools.easyviz`) package, along with some additional numerical functionality.

With Easyviz one can merge several plotting commands into a single one using keyword arguments:

```
plot(t,   u,    'r--o',             # red dashes w/circles
     t_e, u_e, 'b-',                # blue line for exact sol.
     legend=['numerical', 'exact'],
     xlabel='t',
     ylabel='u',
     title='theta=%g, dt=%g' % (theta, dt),
     savefig='%s_%g.png' % (theta2name[theta], dt),
     show=True)
```

The `decay_plot_st.py`[12] file contains such a demo.

By default, Easyviz employs Matplotlib for plotting, but Gnuplot[13] and Grace[14] are viable alternatives:

```
Terminal> python decay_plot_st.py --SCITOOLS_easyviz_backend gnuplot
Terminal> python decay_plot_st.py --SCITOOLS_easyviz_backend grace
```

The actual tool used for creating plots (called *backend*) and numerous other options can be permanently set in SciTool's configuration file.

All the Gnuplot windows are launched without any need to kill one before the next one pops up (as is the case with Matplotlib) and one can press the key 'q' anywhere in a plot window to kill it. Another advantage of Gnuplot is the automatic choice of sensible and distinguishable line types in black-and-white PDF and PostScript files.

For more detailed information on syntax and plotting capabilities, we refer to the Matplotlib [5] and SciTools [7] documentation. The hope is that the programming syntax explained so far suffices for understanding the basic plotting functionality and being able to look up the cited technical documentation.

[11] https://github.com/hplgit/scitools
[12] http://tinyurl.com/ofkw6kc/alg/decay_plot_st.py
[13] http://www.gnuplot.info/
[14] http://plasma-gate.weizmann.ac.il/Grace/

Test your understanding!
Exercise 4.3 asks you to implement a solver for a problem that is slightly different from the one above. You may use the `solver` and `explore` functions explained above as a starting point. Apply the new solver to solve Exercise 4.4.

1.2.12 Memory-Saving Implementation

The computer memory requirements of our implementations so far consist mainly of the `u` and `t` arrays, both of length $N_t + 1$. Also, for the programs that involve array arithmetics, Python needs memory space for storing temporary arrays. For example, computing `I*exp(-a*t)` requires storing the intermediate result `a*t` before the preceding minus sign can be applied. The resulting array is temporarily stored and provided as input to the `exp` function. Regardless of how we implement simple ODE problems, storage requirements are very modest and put no restrictions on how we choose our data structures and algorithms. Nevertheless, when the presented methods are applied to three-dimensional PDE problems, memory storage requirements suddenly become a challenging issue.

Let us briefly elaborate on how large the storage requirements can quickly be in three-dimensional problems. The PDE counterpart to our model problem $u' = -a$ is a diffusion equation $u_t = a\nabla^2 u$ posed on a space-time domain. The discrete representation of this domain may in 3D be a spatial mesh of M^3 points and a time mesh of N_t points. In many applications, it is quite typical that M is at least 100, or even 1000. Storing all the computed u values, like we have done in the programs so far, would demand storing arrays of size up to $M^3 N_t$. This would give a factor of M^3 larger storage demands compared to what was required by our ODE programs. Each real number in the `u` array requires 8 bytes (b) of storage. With $M = 100$ and $N_t = 1000$, there is a storage demand of $(10^3)^3 \cdot 1000 \cdot 8 = 8$ Gb for the solution array. Fortunately, we can usually get rid of the N_t factor, resulting in 8 Mb of storage. Below we explain how this is done (the technique is almost always applied in implementations of PDE problems).

Let us critically evaluate how much we really need to store in the computer's memory for our implementation of the θ method. To compute a new u^{n+1}, all we need is u^n. This implies that the previous $u^{n-1}, u^{n-2}, \ldots, u^0$ values do not need to be stored, although this is convenient for plotting and data analysis in the program. Instead of the `u` array we can work with two variables for real numbers, `u` and `u_1`, representing u^{n+1} and u^n in the algorithm, respectively. At each time level, we update `u` from `u_1` and then set `u_1 = u`, so that the computed u^{n+1} value becomes the "previous" value u^n at the next time level. The downside is that we cannot plot the solution after the simulation is done since only the last two numbers are available. The remedy is to store computed values in a file and use the file for visualizing the solution later.

We have implemented this memory saving idea in the file `decay_memsave.py`[15], which is a slight modification of `decay_plot_mpl.py`[16] program.

[15] http://tinyurl.com/ofkw6kc/alg/decay_memsave.py
[16] http://tinyurl.com/ofkw6kc/alg/decay_plot_mpl.py

The following function demonstrates how we work with the two most recent values of thc unknown:

```
def solver_memsave(I, a, T, dt, theta, filename='sol.dat'):
    """
    Solve u'=-a*u, u(0)=I, for t in (0,T] with steps of dt.
    Minimum use of memory. The solution is stored in a file
    (with name filename) for later plotting.
    """
    dt = float(dt)            # avoid integer division
    Nt = int(round(T/dt))     # no of intervals

    outfile = open(filename, 'w')
    # u: time level n+1, u_1: time level n
    t = 0
    u_1 = I
    outfile.write('%.16E  %.16E\n' % (t, u_1))
    for n in range(1, Nt+1):
        u = (1 - (1-theta)*a*dt)/(1 + theta*dt*a)*u_1
        u_1 = u
        t += dt
        outfile.write('%.16E  %.16E\n' % (t, u))
    outfile.close()
    return u, t
```

This code snippet also serves as a quick introduction to file writing in Python. Reading the data in the file into arrays `t` and `u` is done by the function

```
def read_file(filename='sol.dat'):
    infile = open(filename, 'r')
    u = [];  t = []
    for line in infile:
        words = line.split()
        if len(words) != 2:
            print 'Found more than two numbers on a line!', words
            sys.exit(1)  # abort
        t.append(float(words[0]))
        u.append(float(words[1]))
    return np.array(t), np.array(u)
```

This type of file with numbers in rows and columns is very common, and `numpy` has a function `loadtxt` which loads such tabular data into a two-dimensional array named by the user. Say the name is `data`, the number in row `i` and column `j` is then `data[i,j]`. The whole column number `j` can be extracted by `data[:,j]`. A version of `read_file` using `np.loadtxt` reads

```
def read_file_numpy(filename='sol.dat'):
    data = np.loadtxt(filename)
    t = data[:,0]
    u = data[:,1]
    return t, u
```

The present counterpart to the `explore` function from `decay_plot_mpl.py`[17] must run `solver_memsave` and then load data from file before we can compute the error measure and make the plot:

[17] http://tinyurl.com/ofkw6kc/alg/decay_plot_mpl.py

```
def explore(I, a, T, dt, theta=0.5, makeplot=True):
    filename = 'u.dat'
    u, t = solver_memsave(I, a, T, dt, theta, filename)

    t, u = read_file(filename)
    u_e = u_exact(t, I, a)
    e = u_e - u
    E = sqrt(dt*np.sum(e**2))
    if makeplot:
        figure()
        ...
```

Apart from the internal implementation, where u^n values are stored in a file rather than in an array, `decay_memsave.py` file works exactly as the `decay_plot_mpl.py` file.

1.3 Exercises

Exercise 1.1: Define a mesh function and visualize it

a) Write a function `mesh_function(f, t)` that returns an array with mesh point values $f(t_0), \ldots, f(t_{N_t})$, where `f` is a Python function implementing a mathematical function `f(t)` and $t_0, \ldots, t_{N_t}$ are mesh points stored in the array `t`. Use a loop over the mesh points and compute one mesh function value at the time.
b) Use `mesh_function` to compute the mesh function corresponding to

$$f(t) = \begin{cases} e^{-t}, & 0 \le t \le 3, \\ e^{-3t}, & 3 < t \le 4 \end{cases}$$

Choose a mesh $t_n = n\Delta t$ with $\Delta t = 0.1$. Plot the mesh function.

Filename: `mesh_function`.

Remarks In Sect. 1.2.9 we show how easy it is to compute a mesh function by array arithmetics (or array computing). Using this technique, one could simply implement `mesh_function(f,t)` as `return f(t)`. However, `f(t)` will not work if there are if tests involving `t` inside `f` as is the case in b). Typically, `if t < 3` must have `t < 3` as a boolean expression, but if `t` is array, `t < 3`, is an *array of boolean values*, which is not legal as a boolean expression in an if test. Computing one element at a time as suggested in a) is a way of out of this problem.

We also remark that the function in b) is the solution of $u' = -au$, $u(0) = 1$, for $t \in [0, 4]$, where $a = 1$ for $t \in [0, 3]$ and $a = 3$ for $t \in [3, 4]$.

Problem 1.2: Differentiate a function

Given a mesh function u^n as an array u with u^n values at mesh points $t_n = n\Delta t$, the discrete derivative can be based on centered differences:

$$d^n = [D_{2t}u]^n = \frac{u^{n+1} - u^{n-1}}{2\Delta t}, \quad n = 1, \ldots, N_t - 1 . \tag{1.58}$$

At the end points we use forward and backward differences:

$$d^0 = [D_t^+ u]^n = \frac{u^1 - u^0}{\Delta t},$$

and

$$d^{N_t} = [D_t^- u]^n = \frac{u^{N_t} - u^{N_t-1}}{\Delta t}.$$

a) Write a function `differentiate(u, dt)` that returns the discrete derivative d^n of the mesh function u^n. The parameter `dt` reflects the mesh spacing Δt. Write a corresponding test function `test_differentiate()` for verifying the implementation.

Hint The three differentiation formulas are exact for quadratic polynomials. Use this property to verify the program.

b) A standard implementation of the formula (1.58) is to have a loop over i. For large N_t, such loop may run slowly in Python. A technique for speeding up the computations, called vectorization or array computing, replaces the loop by array operations. To see how this can be done in the present mathematical problem, we define two arrays

$$u^+ = (u^2, u^3, \ldots, u^{N_t}), \quad u^- = (u^0, u^1, \ldots, u^{N_t-2}).$$

The formula (1.58) can now be expressed as

$$(d^1, d^2, \ldots, d^{N_t-1}) = \frac{1}{2\Delta t}(u^+ - u^-).$$

The corresponding Python code reads

```
d[1:-1] = (u[2:] - u[0:-2])/(2*dt)
# or
d[1:N_t] = (u[2:N_t+1] - u[0:N_t-1])/(2*dt)
```

Recall that an array slice `u[1:-1]` contains the elements in `u` starting with index 1 and going all indices up to, but not including, the last one (`-1`).
Use the ideas above to implement a vectorized version of the `differentiate` function without loops. Make a corresponding test function that compares the result with that of `differentiate`.

Filename: `differentiate`.

Problem 1.3: Experiment with divisions
Explain what happens in the following computations, where some are mathematically unexpected:

```
>>> dt = 3
>>> T = 8
>>> Nt = T/dt
>>> Nt
2
>>> theta = 1; a = 1
>>> (1 - (1-theta)*a*dt)/(1 + theta*dt*a)
0
```

Filename: `pyproblems`.

Problem 1.4: Experiment with wrong computations
Consider the `solver` function in the `decay_v1.py`[18] file and the following call:

```
u, t = solver(I=1, a=1, T=7, dt=2, theta=1)
```

The output becomes

```
t= 0.000 u=1
t= 2.000 u=0
t= 4.000 u=0
t= 6.000 u=0
```

Print out the result of all intermediate computations and use `type(v)` to see the object type of the result stored in some variable `v`. Examine the intermediate calculations and explain why `u` is wrong and why we compute up to $t = 6$ only even though we specified $T = 7$.

Filename: `decay_v1_err`.

Problem 1.5: Plot the error function
Solve the problem $u' = -au$, $u(0) = I$, using the Forward Euler, Backward Euler, and Crank–Nicolson schemes. For each scheme, plot the error mesh function $e^n = u_e(t_n) - u^n$ for $\Delta t = 0.1, 0.05, 0.025$, where u_e is the exact solution of the ODE and u^n is the numerical solution at mesh point t_n.

Hint Modify the `decay_plot_mpl.py`[19] code.
Filename: `decay_plot_error`.

Problem 1.6: Change formatting of numbers and debug
The `decay_memsave.py`[20] program writes the time values and solution values to a file which looks like

```
0.0000000000000000E+00  1.0000000000000000E+00
2.0000000000000001E-01  8.3333333333333337E-01
4.0000000000000002E-01  6.9444444444444453E-01
6.0000000000000009E-01  5.7870370370370383E-01
8.0000000000000004E-01  4.8225308641975323E-01
1.0000000000000000E+00  4.0187757201646102E-01
1.2000000000000000E+00  3.3489797668038418E-01
1.3999999999999999E+00  2.7908164723365347E-01
```

[18] http://tinyurl.com/ofkw6kc/alg/decay_v1.py
[19] http://tinyurl.com/ofkw6kc/alg/decay_plot_mpl.py
[20] http://tinyurl.com/ofkw6kc/alg/decay_memsave.py

Modify the file output such that it looks like

```
0.000  1.00000
0.200  0.83333
0.400  0.69444
0.600  0.57870
0.800  0.48225
1.000  0.40188
1.200  0.33490
1.400  0.27908
```

If you have just modified the formatting of numbers in the file, running the modified program

```
Terminal> python decay_memsave_v2.py --T 10 --theta 1 \
          --dt 0.2 --makeplot
```

leads to printing of the message `Bug in the implementation!` in the terminal window. Why?

Filename: `decay_memsave_v2`.

Analysis

2

We address the ODE for exponential decay,

$$u'(t) = -au(t), \quad u(0) = I, \tag{2.1}$$

where a and I are given constants. This problem is solved by the θ-rule finite difference scheme, resulting in the recursive equations

$$u^{n+1} = \frac{1 - (1-\theta)a\Delta t}{1 + \theta a \Delta t} u^n \tag{2.2}$$

for the numerical solution u^{n+1}, which approximates the exact solution u_e at time point t_{n+1}. For constant mesh spacing, which we assume here, $t_{n+1} = (n+1)\Delta t$.

The example programs associated with this chapter are found in the directory `src/analysis`[1].

2.1 Experimental Investigations

We first perform a series of numerical explorations to see how the methods behave as we change the parameters I, a, and Δt in the problem.

2.1.1 Discouraging Numerical Solutions

Choosing $I = 1$, $a = 2$, and running experiments with $\theta = 1, 0.5, 0$ for $\Delta t = 1.25, 0.75, 0.5, 0.1$, gives the results in Figs. 2.1, 2.2, and 2.3.

The characteristics of the displayed curves can be summarized as follows:

- The Backward Euler scheme gives a monotone solution in all cases, lying above the exact curve.
- The Crank–Nicolson scheme gives the most accurate results, but for $\Delta t = 1.25$ the solution oscillates.

[1] http://tinyurl.com/ofkw6kc/analysis

H.P. Langtangen, *Finite Difference Computing with Exponential Decay Models*, Lecture Notes in Computational Science and Engineering 110,
DOI 10.1007/978-3-319-29439-1_2

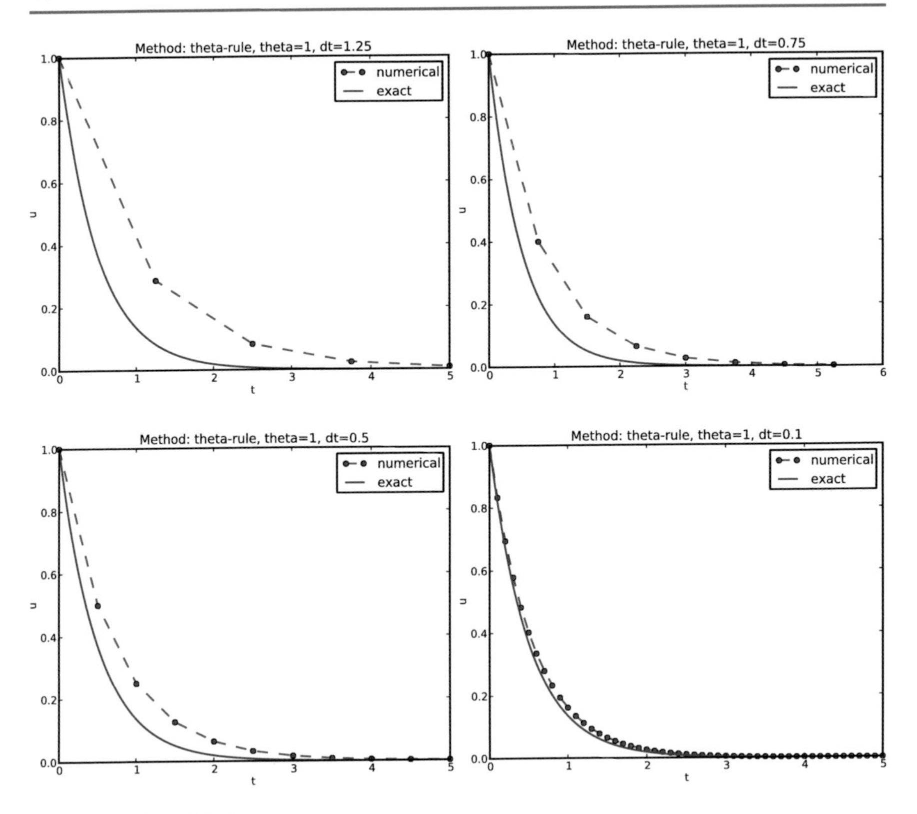

Fig. 2.1 Backward Euler

- The Forward Euler scheme gives a growing, oscillating solution for $\Delta t = 1.25$; a decaying, oscillating solution for $\Delta t = 0.75$; a strange solution $u^n = 0$ for $n \geq 1$ when $\Delta t = 0.5$; and a solution seemingly as accurate as the one by the Backward Euler scheme for $\Delta t = 0.1$, but the curve lies below the exact solution.

Since the exact solution of our model problem is a monotone function, $u(t) = Ie^{-at}$, some of these qualitatively wrong results indeed seem alarming!

Key questions

- Under what circumstances, i.e., values of the input data I, a, and Δt will the Forward Euler and Crank–Nicolson schemes result in undesired oscillatory solutions?
- How does Δt impact the error in the numerical solution?

The first question will be investigated both by numerical experiments and by precise mathematical theory. The theory will help establish general criteria on Δt for avoiding non-physical oscillatory or growing solutions.

For our simple model problem we can answer the second question very precisely, but we will also look at simplified formulas for small Δt and touch upon

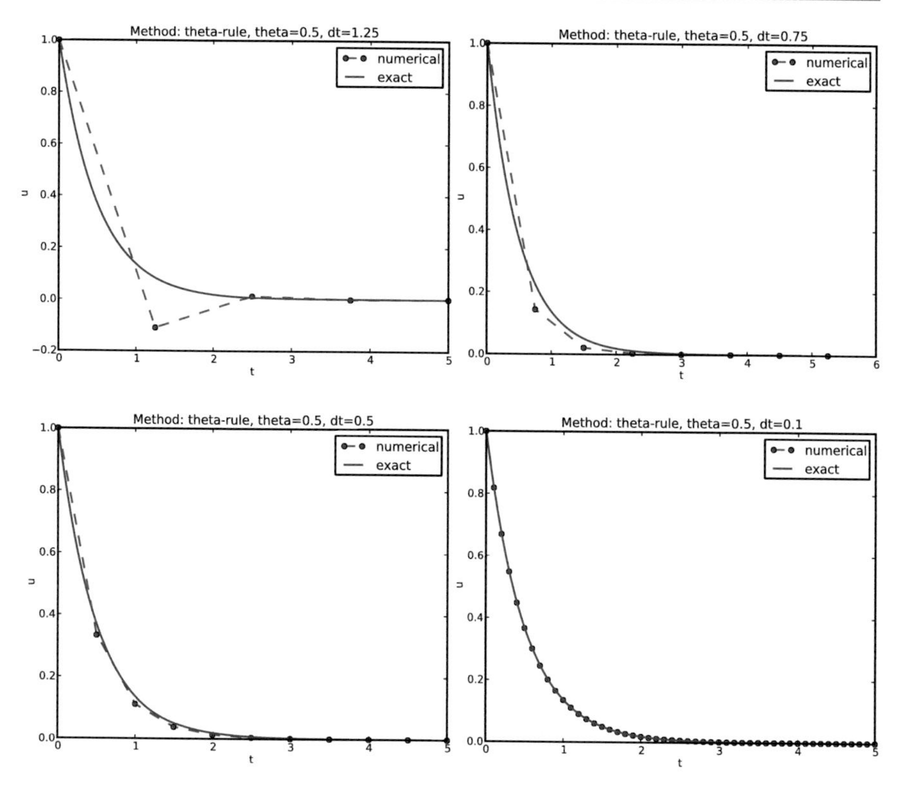

Fig. 2.2 Crank–Nicolson

important concepts such as *convergence rate* and *the order of a scheme*. Other fundamental concepts mentioned are stability, consistency, and convergence.

2.1.2 Detailed Experiments

To address the first question above, we may set up an experiment where we loop over values of I, a, and Δt in our chosen model problem. For each experiment, we flag the solution as oscillatory if

$$u^n > u^{n-1},$$

for some value of n. This seems like a reasonable choice, since we expect u^n to decay with n, but oscillations will make u increase over a time step. Doing some initial experimentation with varying I, a, and Δt, quickly reveals that oscillations are independent of I, but they do depend on a and Δt. We can therefore limit the investigation to vary a and Δt. Based on this observation, we introduce a two-dimensional function $B(a, \Delta t)$ which is 1 if oscillations occur and 0 otherwise. We can visualize B as a contour plot (lines for which $B = \text{const}$). The contour $B = 0.5$ corresponds to the borderline between oscillatory regions with $B = 1$ and monotone regions with $B = 0$ in the $a, \Delta t$ plane.

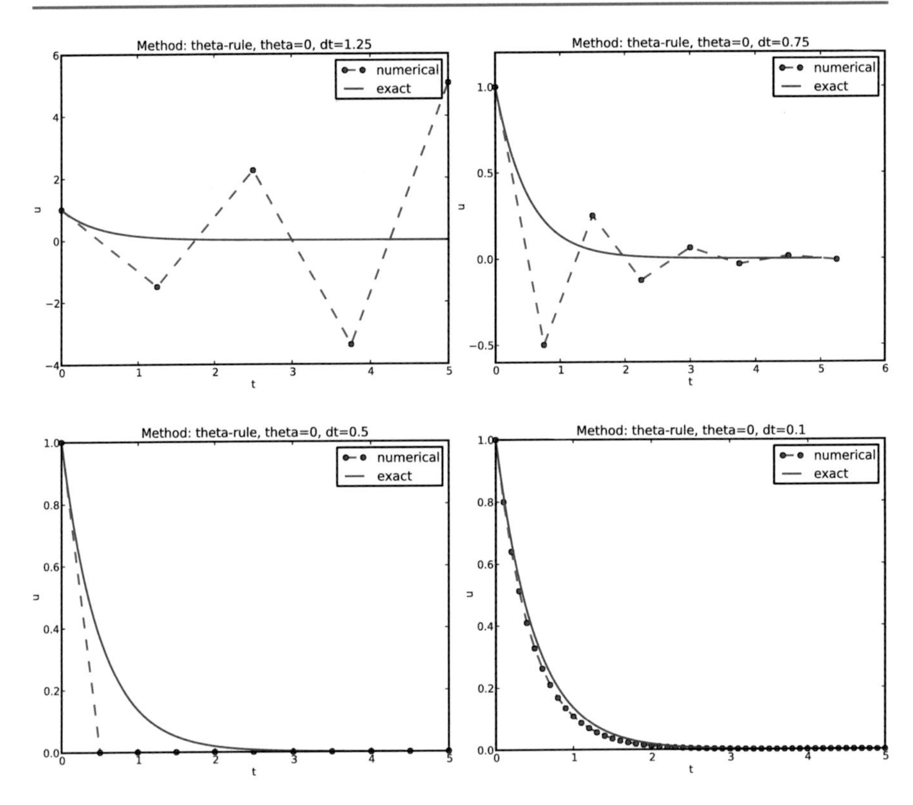

Fig. 2.3 Forward Euler

The B function is defined at discrete a and Δt values. Say we have given P values for a, $a_0, \ldots, a_{P-1}$, and Q values for Δt, $\Delta t_0, \ldots, \Delta t_{Q-1}$. These a_i and Δt_j values, $i = 0, \ldots, P-1$, $j = 0, \ldots, Q-1$, form a rectangular mesh of $P \times Q$ points in the plane spanned by a and Δt. At each point $(a_i, \Delta t_j)$, we associate the corresponding value $B(a_i, \Delta t_j)$, denoted B_{ij}. The B_{ij} values are naturally stored in a two dimensional array. We can thereafter create a plot of the contour line $B_{ij} = 0.5$ dividing the oscillatory and monotone regions. The file `decay_osc_regions.py`[2] given below (`osc_regions` stands for "oscillatory regions") contains all nuts and bolts to produce the $B = 0.5$ line in Figs. 2.4 and 2.5. The oscillatory region is above this line.

```
from decay_mod import solver
import numpy as np
import scitools.std as st

def non_physical_behavior(I, a, T, dt, theta):
    """
    Given lists/arrays a and dt, and numbers I, dt, and theta,
    make a two-dimensional contour line B=0.5, where B=1>0.5
    means oscillatory (unstable) solution, and B=0<0.5 means
    monotone solution of u'=-au.
    """
```

[2] http://tinyurl.com/ofkw6kc/analysis/decay_osc_regions.py

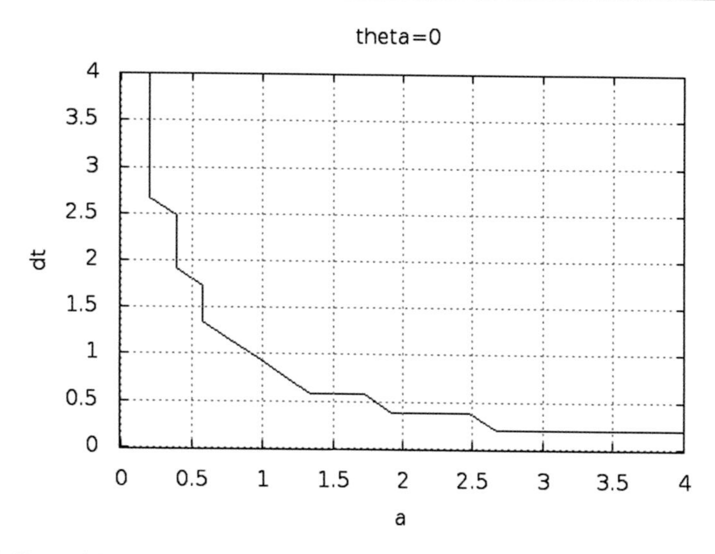

Fig. 2.4 Forward Euler scheme: oscillatory solutions occur for points above the curve

```
    a = np.asarray(a); dt = np.asarray(dt)  # must be arrays
    B = np.zeros((len(a), len(dt)))         # results
    for i in range(len(a)):
        for j in range(len(dt)):
            u, t = solver(I, a[i], T, dt[j], theta)
            # Does u have the right monotone decay properties?
            correct_qualitative_behavior = True
            for n in range(1, len(u)):
                if u[n] > u[n-1]:  # Not decaying?
                    correct_qualitative_behavior = False
                    break  # Jump out of loop
            B[i,j] = float(correct_qualitative_behavior)
    a_, dt_ = st.ndgrid(a, dt)  # make mesh of a and dt values
    st.contour(a_, dt_, B, 1)
    st.grid('on')
    st.title('theta=%g' % theta)
    st.xlabel('a'); st.ylabel('dt')
    st.savefig('osc_region_theta_%s.png' % theta)
    st.savefig('osc_region_theta_%s.pdf' % theta)

non_physical_behavior(
    I=1,
    a=np.linspace(0.01, 4, 22),
    dt=np.linspace(0.01, 4, 22),
    T=6,
    theta=0.5)
```

By looking at the curves in the figures one may guess that $a\Delta t$ must be less than a critical limit to avoid the undesired oscillations. This limit seems to be about 2 for Crank–Nicolson and 1 for Forward Euler. We shall now establish a precise mathematical analysis of the discrete model that can explain the observations in our numerical experiments.

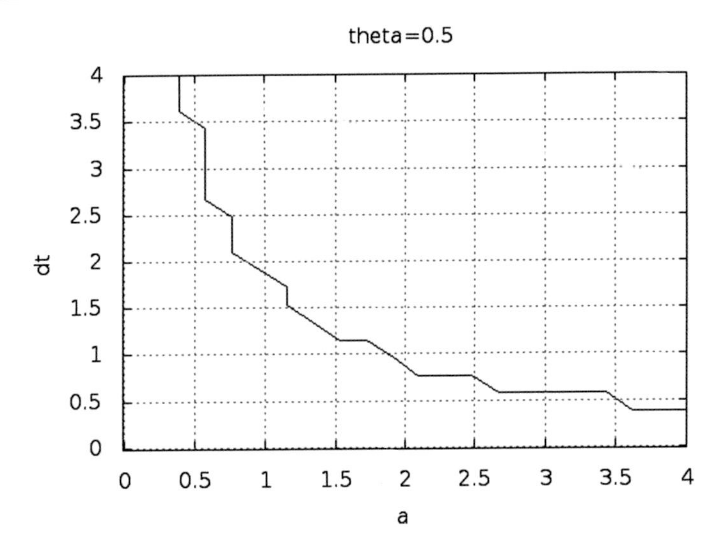

Fig. 2.5 Crank–Nicolson scheme: oscillatory solutions occur for points above the curve

2.2 Stability

The goal now is to understand the results in the previous section. To this end, we shall investigate the properties of the mathematical formula for the solution of the equations arising from the finite difference methods.

2.2.1 Exact Numerical Solution

Starting with $u^0 = I$, the simple recursion (2.2) can be applied repeatedly n times, with the result that

$$u^n = IA^n, \quad A = \frac{1 - (1-\theta)a\Delta t}{1 + \theta a\Delta t}\,. \tag{2.3}$$

Solving difference equations

Difference equations where all terms are linear in u^{n+1}, u^n, and maybe u^{n-1}, u^{n-2}, etc., are called *homogeneous, linear* difference equations, and their solutions are generally of the form $u^n = A^n$, where A is a constant to be determined. Inserting this expression in the difference equation and dividing by A^{n+1} gives a polynomial equation in A. In the present case we get

$$A = \frac{1 - (1-\theta)a\Delta t}{1 + \theta a\Delta t}\,.$$

This is a solution technique of wider applicability than repeated use of the recursion (2.2).

Regardless of the solution approach, we have obtained a formula for u^n. This formula can explain everything we see in the figures above, but it also gives us a more general insight into accuracy and stability properties of the three schemes.

Since u^n is a factor A raised to an integer power n, we realize that $A < 0$ will imply $u^n < 0$ for odd n and $u^n > 0$ for even n. That is, the solution oscillates between the mesh points. We have oscillations due to $A < 0$ when

$$(1-\theta)a\Delta t > 1\,. \tag{2.4}$$

Since $A > 0$ is a requirement for having a numerical solution with the same basic property (monotonicity) as the exact solution, we may say that $A > 0$ is a *stability criterion*. Expressed in terms of Δt the stability criterion reads

$$\Delta t < \frac{1}{(1-\theta)a}\,. \tag{2.5}$$

The Backward Euler scheme is always stable since $A < 0$ is impossible for $\theta = 1$, while non-oscillating solutions for Forward Euler and Crank–Nicolson demand $\Delta t \leq 1/a$ and $\Delta t \leq 2/a$, respectively. The relation between Δt and a look reasonable: a larger a means faster decay and hence a need for smaller time steps.

Looking at the upper left plot in Fig. 2.3, we see that $\Delta t = 1.25$, and remembering that $a = 2$ in these experiments, A can be calculated to be -1.5, so the Forward Euler solution becomes $u^n = (-1.5)^n$ ($I = 1$). This solution oscillates *and* grows. The upper right plot has $a\Delta t = 2 \cdot 0.75 = 1.5$, so $A = -0.5$, and $u^n = (-0.5)^n$ decays but oscillates. The lower left plot is a peculiar case where the Forward Euler scheme produces a solution that is stuck on the t axis. Now we can understand why this is so, because $a\Delta t = 2 \cdot 0.5 = 1$, which gives $A = 0$, and therefore $u^n = 0$ for $n \geq 1$. The decaying oscillations in the Crank–Nicolson scheme in the upper left plot in Fig. 2.2 for $\Delta t = 1.25$ are easily explained by the fact that $A \approx -0.11 < 0$.

2.2.2 Stability Properties Derived from the Amplification Factor

The factor A is called the *amplification factor* since the solution at a new time level is the solution at the previous time level amplified by a factor A. For a decay process, we must obviously have $|A| \leq 1$, which is fulfilled for all Δt if $\theta \geq 1/2$. Arbitrarily large values of u can be generated when $|A| > 1$ and n is large enough. The numerical solution is in such cases totally irrelevant to an ODE modeling decay processes! To avoid this situation, we must demand $|A| \leq 1$ also for $\theta < 1/2$, which implies

$$\Delta t \leq \frac{2}{(1-2\theta)a}, \tag{2.6}$$

For example, Δt must not exceed $2/a$ when computing with the Forward Euler scheme.

Stability properties

We may summarize the stability investigations as follows:

1. The Forward Euler method is a *conditionally stable* scheme because it requires $\Delta t < 2/a$ for avoiding growing solutions and $\Delta t < 1/a$ for avoiding oscillatory solutions.
2. The Crank–Nicolson is *unconditionally stable* with respect to growing solutions, while it is conditionally stable with the criterion $\Delta t < 2/a$ for avoiding oscillatory solutions.
3. The Backward Euler method is unconditionally stable with respect to growing and oscillatory solutions – any Δt will work.

Much literature on ODEs speaks about L-stable and A-stable methods. In our case A-stable methods ensures non-growing solutions, while L-stable methods also avoids oscillatory solutions.

2.3 Accuracy

While stability concerns the qualitative properties of the numerical solution, it remains to investigate the quantitative properties to see exactly how large the numerical errors are.

2.3.1 Visual Comparison of Amplification Factors

After establishing how A impacts the qualitative features of the solution, we shall now look more into how well the numerical amplification factor approximates the exact one. The exact solution reads $u(t) = Ie^{-at}$, which can be rewritten as

$$u_{\mathrm{e}}(t_n) = Ie^{-an\Delta t} = I(e^{-a\Delta t})^n\,. \tag{2.7}$$

From this formula we see that the exact amplification factor is

$$A_{\mathrm{e}} = e^{-a\Delta t}\,. \tag{2.8}$$

We see from all of our analysis that the exact and numerical amplification factors depend on a and Δt through the dimensionless product $a\Delta t$: whenever there is a Δt in the analysis, there is always an associated a parameter. Therefore, it is convenient to introduce a symbol for this product, $p = a\Delta t$, and view A and A_{e} as functions of p. Figure 2.6 shows these functions. The two amplification factors are clearly closest for the Crank–Nicolson method, but that method has the unfortunate oscillatory behavior when $p > 2$.

Significance of the $p = a\Delta t$ parameter

The key parameter for numerical performance of a scheme is in this model problem $p = a\Delta t$. This is a *dimensionless number* (a has dimension 1/s and Δt

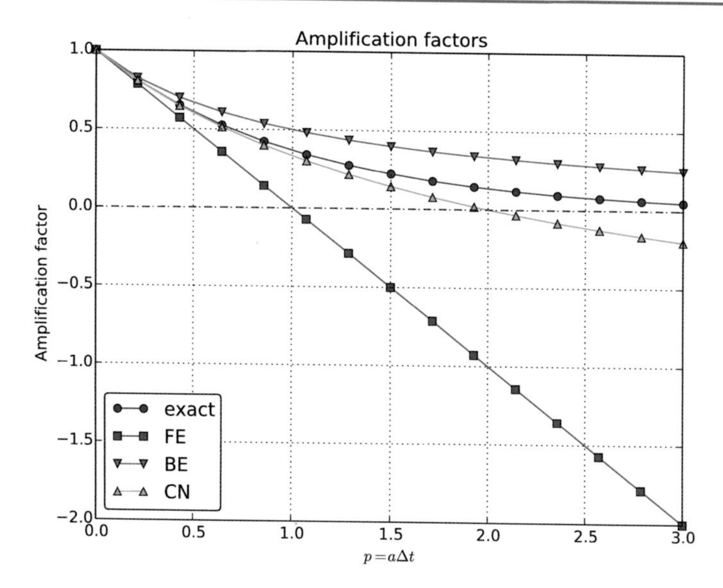

Fig. 2.6 Comparison of amplification factors

has dimension s) reflecting how the discretization parameter plays together with a physical parameter in the problem.

One can bring the present model problem on dimensionless form through a process called scaling. The scaled modeled has a modified time $\bar{t} = at$ and modified response $\bar{u} = u/I$ such that the model reads $d\bar{u}/d\bar{t} = -\bar{u}$, $\bar{u}(0) = 1$. Analyzing this model, where there are no physical parameters, we find that $\Delta\bar{t}$ is the key parameter for numerical performance. In the unscaled model, this corresponds to $\Delta\bar{t} = a\Delta t$.

It is common that the numerical performance of methods for solving ordinary and partial differential equations is governed by dimensionless parameters that combine mesh sizes with physical parameters.

2.3.2 Series Expansion of Amplification Factors

As an alternative to the visual understanding inherent in Fig. 2.6, there is a strong tradition in numerical analysis to establish formulas for approximation errors when the discretization parameter, here Δt, becomes small. In the present case, we let p be our small discretization parameter, and it makes sense to simplify the expressions for A and A_{e} by using Taylor polynomials around $p = 0$. The Taylor polynomials are accurate for small p and greatly simplify the comparison of the analytical expressions since we then can compare polynomials, term by term.

Calculating the Taylor series for A_{e} is easily done by hand, but the three versions of A for $\theta = 0, 1, \frac{1}{2}$ lead to more cumbersome calculations. Nowadays, analytical computations can benefit greatly by symbolic computer algebra software. The Python package `sympy` represents a powerful computer algebra system, not yet as sophisticated as the famous Maple and Mathematica systems, but it is free and very easy to integrate with our numerical computations in Python.

When using sympy, it is convenient to enter an interactive Python shell where the results of expressions and statements can be shown immediately. Here is a simple example. We strongly recommend to use isympy (or ipython) for such interactive sessions.

Let us illustrate sympy with a standard Python shell syntax (>>> prompt) to compute a Taylor polynomial approximation to e^{-p}:

```
>>> from sympy import *
>>> # Create p as a mathematical symbol with name 'p'
>>> p = Symbols('p')
>>> # Create a mathematical expression with p
>>> A_e = exp(-p)
>>>
>>> # Find the first 6 terms of the Taylor series of A_e
>>> A_e.series(p, 0, 6)
1 + (1/2)*p**2 - p - 1/6*p**3 - 1/120*p**5 + (1/24)*p**4 + O(p**6)
```

Lines with >>> represent input lines, whereas without this prompt represent the result of the previous command (note that isympy and ipython apply other prompts, but in this text we always apply >>> for interactive Python computing). Apart from the order of the powers, the computed formula is easily recognized as the beginning of the Taylor series for e^{-p}.

Let us define the numerical amplification factor where p and θ enter the formula as symbols:

```
>>> theta = Symbol('theta')
>>> A = (1-(1-theta)*p)/(1+theta*p)
```

To work with the factor for the Backward Euler scheme we can substitute the value 1 for theta:

```
>>> A.subs(theta, 1)
1/(1 + p)
```

Similarly, we can substitute theta by $1/2$ for Crank–Nicolson, preferably using an exact rational representation of $1/2$ in sympy:

```
>>> half = Rational(1,2)
>>> A.subs(theta, half)
1/(1 + (1/2)*p)*(1 - 1/2*p)
```

The Taylor series of the amplification factor for the Crank–Nicolson scheme can be computed as

```
>>> A.subs(theta, half).series(p, 0, 4)
1 + (1/2)*p**2 - p - 1/4*p**3 + O(p**4)
```

We are now in a position to compare Taylor series:

```
>>> FE = A_e.series(p, 0, 4) - A.subs(theta, 0).series(p, 0, 4)
>>> BE = A_e.series(p, 0, 4) - A.subs(theta, 1).series(p, 0, 4)
>>> CN = A_e.series(p, 0, 4) - A.subs(theta, half).series(p, 0, 4 )
>>> FE
(1/2)*p**2 - 1/6*p**3 + O(p**4)
>>> BE
-1/2*p**2 + (5/6)*p**3 + O(p**4)
>>> CN
(1/12)*p**3 + O(p**4)
```

From these expressions we see that the error $A - A_e \sim \mathcal{O}(p^2)$ for the Forward and Backward Euler schemes, while $A - A_e \sim \mathcal{O}(p^3)$ for the Crank–Nicolson scheme. The notation $\mathcal{O}(p^m)$ here means a polynomial in p where p^m is the term of lowest-degree, and consequently the term that dominates the expression for $p < 0$. We call this the *leading order term*. As $p \to 0$, the leading order term clearly dominates over the higher-order terms (think of $p = 0.01$: p is a hundred times larger than p^2).

Now, a is a given parameter in the problem, while Δt is what we can vary. Not surprisingly, the error expressions are usually written in terms Δt. We then have

$$A - A_e = \begin{cases} \mathcal{O}(\Delta t^2), & \text{Forward and Backward Euler,} \\ \mathcal{O}(\Delta t^3), & \text{Crank–Nicolson} \end{cases} \tag{2.9}$$

We say that the Crank–Nicolson scheme has an error in the amplification factor of order Δt^3, while the two other schemes are of order Δt^2 in the same quantity.

What is the significance of the order expression? If we halve Δt, the error in amplification factor at a time level will be reduced by a factor of 4 in the Forward and Backward Euler schemes, and by a factor of 8 in the Crank–Nicolson scheme. That is, as we reduce Δt to obtain more accurate results, the Crank–Nicolson scheme reduces the error more efficiently than the other schemes.

2.3.3 The Ratio of Numerical and Exact Amplification Factors

An alternative comparison of the schemes is provided by looking at the ratio A/A_e, or the error $1 - A/A_e$ in this ratio:

```
>>> FE = 1 - (A.subs(theta, 0)/A_e).series(p, 0, 4)
>>> BE = 1 - (A.subs(theta, 1)/A_e).series(p, 0, 4)
>>> CN = 1 - (A.subs(theta, half)/A_e).series(p, 0, 4)
>>> FE
(1/2)*p**2 + (1/3)*p**3 + O(p**4)
>>> BE
-1/2*p**2 + (1/3)*p**3 + O(p**4)
>>> CN
(1/12)*p**3 + O(p**4)
```

The leading-order terms have the same powers as in the analysis of $A - A_e$.

2.3.4 The Global Error at a Point

The error in the amplification factor reflects the error when progressing from time level t_n to t_{n-1} only. That is, we disregard the error already present in the solution at t_{n-1}. The real error at a point, however, depends on the error development over all previous time steps. This error, $e^n = u^n - u_e(t_n)$, is known as the *global error*. We may look at u^n for some n and Taylor expand the mathematical expressions as functions of $p = a\Delta t$ to get a simple expression for the global error (for small p). Continuing the sympy expression from previous section, we can write

```
>>> n = Symbol('n')
>>> u_e = exp(-p*n)
>>> u_n = A**n
>>> FE = u_e.series(p, 0, 4) - u_n.subs(theta, 0).series(p, 0, 4)
>>> BE = u_e.series(p, 0, 4) - u_n.subs(theta, 1).series(p, 0, 4)
>>> CN = u_e.series(p, 0, 4) - u_n.subs(theta, half).series(p, 0, 4)
>>> FE
(1/2)*n*p**2 - 1/2*n**2*p**3 + (1/3)*n*p**3 + O(p**4)
>>> BE
(1/2)*n**2*p**3 - 1/2*n*p**2 + (1/3)*n*p**3 + O(p**4)
>>> CN
(1/12)*n*p**3 + O(p**4)
```

Note that sympy does not sort the polynomial terms in the output, so p^3 appears before p^2 in the output of BE.

For a fixed time t, the parameter n in these expressions increases as $p \rightarrow 0$ since $t = n\Delta t =$ const and hence n must increase like Δt^{-1}. With n substituted by $t/\Delta t$ in the leading-order error terms, these become

$$e^n = \frac{1}{2}np^2 = \frac{1}{2}ta^2\Delta t, \qquad \text{Forward Euler} \tag{2.10}$$

$$e^n = -\frac{1}{2}np^2 = -\frac{1}{2}ta^2\Delta t, \qquad \text{Backward Euler} \tag{2.11}$$

$$e^n = \frac{1}{12}np^3 = \frac{1}{12}ta^3\Delta t^2, \qquad \text{Crank–Nicolson} \tag{2.12}$$

The global error is therefore of second order (in Δt) for the Crank–Nicolson scheme and of first order for the other two schemes.

Convergence

When the global error $e^n \rightarrow 0$ as $\Delta t \rightarrow 0$, we say that the scheme is *convergent*. It means that the numerical solution approaches the exact solution as the mesh is refined, and this is a much desired property of a numerical method.

2.3.5 Integrated Error

It is common to study the norm of the numerical error, as explained in detail in Sect. 1.2.10. The L^2 norm of the error can be computed by treating e^n as a function of t in sympy and performing symbolic integration. From now on we shall do `import sympy as sym` and prefix all functions in sympy by sym to explicitly notify ourselves that the functions are from `sympy`. This is particularly advantageous when we use mathematical functions like `sin`: `sym.sin` is for symbolic expressions, while `sin` from `numpy` or `math` is for numerical computation. For the Forward Euler scheme we have

```
import sympy as sym
p, n, a, dt, t, T, theta = sym.symbols('p n a dt t T theta')
A = (1-(1-theta)*p)/(1+theta*p)
u_e = sym.exp(-p*n)
u_n = A**n
error = u_e.series(p, 0, 4) - u_n.subs(theta, 0).series(p, 0, 4)
# Introduce t and dt instead of n and p
error = error.subs('n', 't/dt').subs(p, 'a*dt')
error = error.as_leading_term(dt) # study only the first term
print error
error_L2 = sym.sqrt(sym.integrate(error**2, (t, 0, T)))
print 'L2 error:', sym.simplify(error_error_L2)
```

The output reads

```
sqrt(30)*sqrt(T**3*a**4*dt**2*(6*T**2*a**2 - 15*T*a + 10))/60
```

which means that the L^2 error behaves like $a^2\Delta t$.

Strictly speaking, the numerical error is only defined at the mesh points so it makes most sense to compute the ℓ^2 error

$$||e^n||_{\ell^2} = \sqrt{\Delta t\sum_{n=0}^{N_t}(u_e(t_n) - u^n)^2}\,.$$

We have obtained an exact analytical expression for the error at $t = t_n$, but here we use the leading-order error term only since we are mostly interested in how the error behaves as a polynomial in Δt or p, and then the leading order term will dominate. For the Forward Euler scheme, $u_e(t_n) - u^n \approx \frac{1}{2}np^2$, and we have

$$||e^n||^2_{\ell^2} = \Delta t\sum_{n=0}^{N_t}\frac{1}{4}n^2p^4 = \Delta t\frac{1}{4}p^4\sum_{n=0}^{N_t}n^2\,.$$

Now, $\sum_{n=0}^{N_t} n^2 \approx \frac{1}{3}N_t^3$. Using this approximation, setting $N_t = T/\Delta t$, and taking the square root gives the expression

$$||e^n||_{\ell^2} = \frac{1}{2}\sqrt{\frac{T^3}{3}}a^2\Delta t\,. \tag{2.13}$$

Calculations for the Backward Euler scheme are very similar and provide the same result, while the Crank–Nicolson scheme leads to

$$||e^n||_{\ell^2} = \frac{1}{12}\sqrt{\frac{T^3}{3}}a^3\Delta t^2\,. \tag{2.14}$$

Summary of errors

Both the global point-wise errors (2.10)–(2.12) and their time-integrated versions (2.13) and (2.14) show that

- the Crank–Nicolson scheme is of second order in Δt, and
- the Forward Euler and Backward Euler schemes are of first order in Δt.

2.3.6 Truncation Error

The truncation error is a very frequently used error measure for finite difference methods. It is defined as *the error in the difference equation that arises when inserting the exact solution*. Contrary to many other error measures, e.g., the true error $e^n = u_e(t_n) - u^n$, the truncation error is a quantity that is easily computable.

Before reading on, it is wise to review Sect. 1.1.7 on how Taylor polynomials were used to derive finite differences and quantify the error in the formulas. Very similar reasoning, and almost identical mathematical details, will be carried out below, but in a slightly different context. Now, the focus is on the error when solving a differential equation, while in Sect. 1.1.7 we derived errors for a finite difference formula. These errors are tightly connected in the present model problem.

Let us illustrate the calculation of the truncation error for the Forward Euler scheme. We start with the difference equation on operator form,

$$[D_t^+ u = -au]^n,$$

which is the short form for

$$\frac{u^{n+1} - u^n}{\Delta t} = -au^n .$$

The idea is to see how well the exact solution $u_e(t)$ fulfills this equation. Since $u_e(t)$ in general will not obey the discrete equation, we get an error in the discrete equation. This error is called a *residual*, denoted here by R^n:

$$R^n = \frac{u_e(t_{n+1}) - u_e(t_n)}{\Delta t} + au_e(t_n) . \tag{2.15}$$

The residual is defined at each mesh point and is therefore a mesh function with a superscript n.

The interesting feature of R^n is to see how it depends on the discretization parameter Δt. The tool for reaching this goal is to Taylor expand u_e around the point where the difference equation is supposed to hold, here $t = t_n$. We have that

$$u_e(t_{n+1}) = u_e(t_n) + u_e'(t_n)\Delta t + \frac{1}{2}u_e''(t_n)\Delta t^2 + \cdots ,$$

which may be used to reformulate the fraction in (2.15) so that

$$R^n = u_e'(t_n) + \frac{1}{2}u_e''(t_n)\Delta t + \ldots + au_e(t_n).$$

Now, u_e fulfills the ODE $u_e' = -au_e$, which means that the first and last term cancel and we have

$$R^n = \frac{1}{2}u_e''(t_n)\Delta t + \mathcal{O}(\Delta t^2).$$

This R^n is the *truncation error*, which for the Forward Euler is seen to be of first order in Δt as $\Delta \rightarrow 0$.

The above procedure can be repeated for the Backward Euler and the Crank–Nicolson schemes. We start with the scheme in operator notation, write it out in detail, Taylor expand u_e around the point $\tilde{t}$ at which the difference equation is defined, collect terms that correspond to the ODE (here $u_e' + au_e$), and identify the remaining terms as the residual R, which is the truncation error. The Backward Euler scheme leads to

$$R^n \approx -\frac{1}{2}u_e''(t_n)\Delta t,$$

while the Crank–Nicolson scheme gives

$$R^{n+\frac{1}{2}} \approx \frac{1}{24}u_e'''(t_{n+\frac{1}{2}})\Delta t^2,$$

when $\Delta t \rightarrow 0$.

The *order* r of a finite difference scheme is often defined through the leading term Δt^r in the truncation error. The above expressions point out that the Forward and Backward Euler schemes are of first order, while Crank–Nicolson is of second order. We have looked at other error measures in other sections, like the error in amplification factor and the error $e^n = u_e(t_n) - u^n$, and expressed these error measures in terms of Δt to see the order of the method. Normally, calculating the truncation error is more straightforward than deriving the expressions for other error measures and therefore the easiest way to establish the order of a scheme.

2.3.7 Consistency, Stability, and Convergence

Three fundamental concepts when solving differential equations by numerical methods are consistency, stability, and convergence. We shall briefly touch upon these concepts below in the context of the present model problem.

Consistency means that the error in the difference equation, measured through the truncation error, goes to zero as $\Delta t \rightarrow 0$. Since the truncation error tells how well the exact solution fulfills the difference equation, and the exact solution fulfills the differential equation, consistency ensures that the difference equation approaches the differential equation in the limit. The expressions for the truncation errors in the previous section are all proportional to Δt or Δt^2, hence they vanish as $\Delta t \rightarrow 0$, and all the schemes are consistent. Lack of consistency implies that we actually solve some other differential equation in the limit $\Delta t \rightarrow 0$ than we aim at.

Stability means that the numerical solution exhibits the same qualitative properties as the exact solution. This is obviously a feature we want the numerical solution to have. In the present exponential decay model, the exact solution is monotone and decaying. An increasing numerical solution is not in accordance with the decaying nature of the exact solution and hence unstable. We can also say that an oscillating numerical solution lacks the property of monotonicity of the exact solution and is also unstable. We have seen that the Backward Euler scheme always leads to monotone and decaying solutions, regardless of Δt, and is hence stable. The Forward Euler scheme can lead to increasing solutions and oscillating solutions if Δt is too large and is therefore unstable unless Δt is sufficiently small. The Crank–Nicolson can never lead to increasing solutions and has no problem to fulfill that stability property, but it can produce oscillating solutions and is unstable in that sense, unless Δt is sufficiently small.

Convergence implies that the global (true) error mesh function $e^n = u_e(t_n) - u^n \rightarrow 0$ as $\Delta t \rightarrow 0$. This is really what we want: the numerical solution gets as close to the exact solution as we request by having a sufficiently fine mesh.

Convergence is hard to establish theoretically, except in quite simple problems like the present one. Stability and consistency are much easier to calculate. A major breakthrough in the understanding of numerical methods for differential equations came in 1956 when Lax and Richtmeyer established equivalence between convergence on one hand and consistency and stability on the other (the Lax equivalence theorem[3]). In practice it meant that one can first establish that a method is stable and consistent, and then it is automatically convergent (which is much harder to establish). The result holds for linear problems only, and in the world of nonlinear differential equations the relations between consistency, stability, and convergence are much more complicated.

We have seen in the previous analysis that the Forward Euler, Backward Euler, and Crank–Nicolson schemes are convergent ($e^n \rightarrow 0$), that they are consistent ($R^n \rightarrow 0$), and that they are stable under certain conditions on the size of Δt. We have also derived explicit mathematical expressions for e^n, the truncation error, and the stability criteria.

2.4 Various Types of Errors in a Differential Equation Model

So far we have been concerned with one type of error, namely the discretization error committed by replacing the differential equation problem by a recursive set of difference equations. There are, however, other types of errors that must be considered too. We can classify errors into four groups:

1. model errors: how wrong is the ODE model?
2. data errors: how wrong are the input parameters?
3. discretization errors: how wrong is the numerical method?
4. rounding errors: how wrong is the computer arithmetics?

Below, we shall briefly describe and illustrate these four types of errors. Each of the errors deserve its own chapter, at least, so the treatment here is superficial to

[3] http://en.wikipedia.org/wiki/Lax_equivalence_theorem

give some indication about the nature of size of the errors in a specific case. Some of the required computer codes quickly become more advanced than in the rest of the book, but we include to code to document all the details that lie behind the investigations of the errors.

2.4.1 Model Errors

Any mathematical modeling like $u' = -au$, $u(0) = I$, is just an approximate description of a real-world phenomenon. How good this approximation is can be determined by comparing physical experiments with what the model predicts. This is the topic of *validation* and is obviously an essential part of mathematical modeling. One difficulty with validation is that we need to estimate the parameters in the model, and this brings in data errors. Quantifying data errors is challenging, and a frequently used method is to *tune* the parameters in the model to make model predictions as close as possible to the experiments. That is, we do not attempt to measure or estimate all input parameters, but instead find values that "make the model good". Another difficulty is that the response in experiments also contains errors due to measurement techniques.

Let us try to quantify model errors in a very simple example involving $u' = -au$, $u(0) = I$, with constant a. Suppose a more accurate model has a as a function of time rather than a constant. Here we take $a(t)$ as a simple linear function: $a + pt$. Obviously, u with $p > 0$ will go faster to zero with time than a constant a.

The solution of

$$u' = (a + pt)u, \quad u(0) = I,$$

can be shown (see below) to be

$$u(t) = Ie^{-t\left(a+\frac{1}{2}pt\right)}.$$

Let a Python function `true_model(t, I, a, p)` implement the above $u(t)$ and let the solution of our primary ODE $u' = -au$ be available as the function `model(t, I, a)`. We can now make some plots of the two models and the error for some values of p. Figure 2.7 displays `model` versus `true_model` for $p = 0.01, 0.1, 1$, while Fig. 2.8 shows the difference between the two models as a function of t for the same p values.

The code that was used to produce the plots looks like

```
from numpy import linspace, exp
from matplotlib.pyplot import \
    plot, show, xlabel, ylabel, legend, savefig, figure, title

def model_errors():
    p_values = [0.01, 0.1, 1]
    a = 1
    I = 1
    t = linspace(0, 4, 101)
    legends = []
    # Work with figure(1) for the discrepancy and figure(2+i)
    # for plotting the model and the true model for p value no i
    for i, p in enumerate(p_values):
        u = model(t, I, a)
```

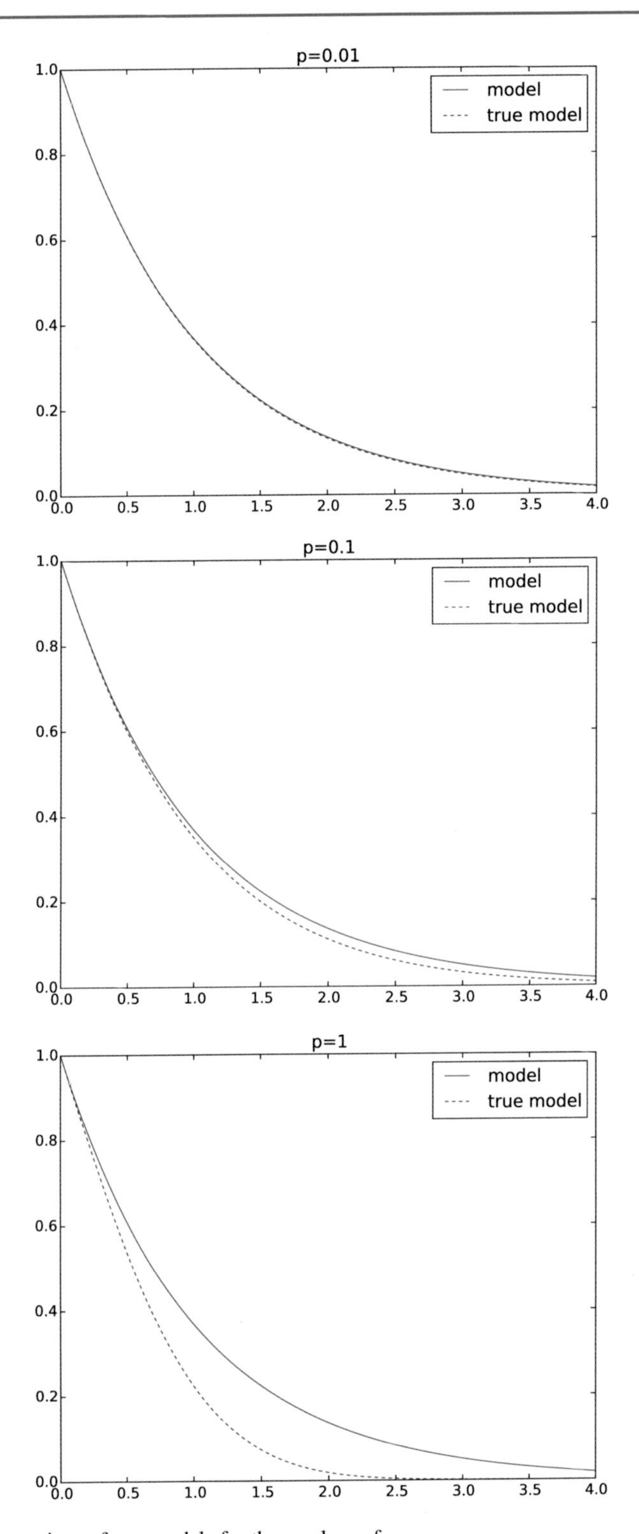

Fig. 2.7 Comparison of two models for three values of p

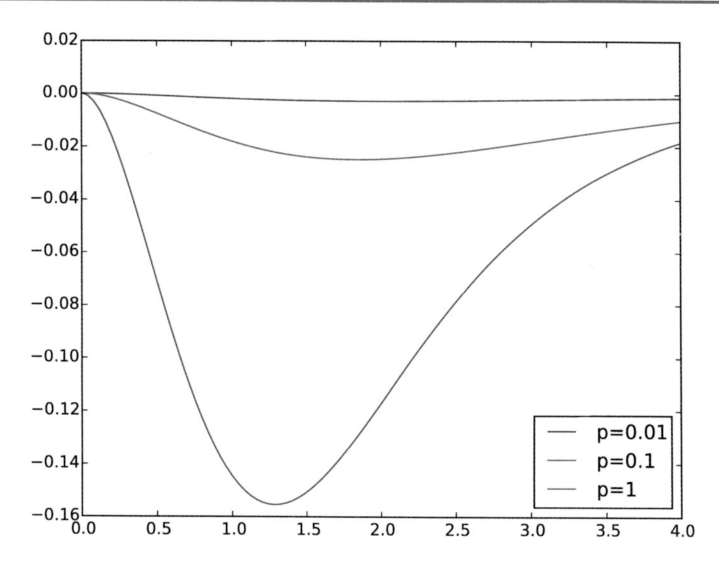

Fig. 2.8 Discrepancy of Comparison of two models for three values of p

```
        u_true = true_model(t, I, a, p)
        discrepancy = u_true - u
        figure(1)
        plot(t, discrepancy)
        figure(2+i)
        plot(t, u, 'r-', t, u_true, 'b--')
        legends.append('p=%g' % p)
    figure(1)
    legend(legends, loc='lower right')
    savefig('tmp1.png'); savefig('tmp1.pdf')
    for i, p in enumerate(p_values):
        figure(2+i)
        legend(['model', 'true model'])
        title('p=%g' % p)
        savefig('tmp%d.png' % (2+i)); savefig('tmp%d.pdf' % (2+i))
```

To derive the analytical solution of the model $u' = -(a + pt)u$, $u(0) = I$, we can use SymPy and the code below. This is somewhat advanced SymPy use for a newbie, but serves to illustrate the possibilities to solve differential equations by symbolic software.

```
def derive_true_solution():
    import sympy as sym
    u = sym.symbols('u', cls=sym.Function)  # function u(t)
    t, a, p, I = sym.symbols('t a p I', real=True)

    def ode(u, t, a, p):
        """Define ODE: u' = (a + p*t)*u. Return residual."""
        return sym.diff(u, t) + (a + p*t)*u

    eq = ode(u(t), t, a, p)
    s = sym.dsolve(eq)
    # s is sym.Eq object u(t) == expression, we want u = expression,
    # so grab the right-hand side of the equality (Eq obj.)
    u = s.rhs
    print u
```

```
    # u contains C1, replace it with a symbol we can fit to
    # the initial condition
    C1 = sym.symbols('C1', real=True)
    u = u.subs('C1', C1)
    print u
    # Initial condition equation
    eq = u.subs(t, 0) - I
    s = sym.solve(eq, C1)  # solve eq wrt C1
    print s
    # s is a list s[0] = ...
    # Replace C1 in u by the solution
    u = u.subs(C1, s[0])
    print 'u:', u
    print sym.latex(u)  # latex formula for reports

    # Consistency check: u must fulfill ODE and initial condition
    print 'ODE is fulfilled:', sym.simplify(ode(u, t, a, p))
    print 'u(0)-I:', sym.simplify(u.subs(t, 0) - I)

    # Convert u expression to Python numerical function
    # (modules='numpy' allows numpy arrays as arguments,
    # we want this for t)
    u_func = sym.lambdify([t, I, a, p], u, modules='numpy')
    return u_func

true_model = derive_true_solution()
```

2.4.2 Data Errors

By "data" we mean all the input parameters to a model, in our case I and a. The values of these may contain errors, or at least uncertainty. Suppose I and a are measured from some physical experiments. Ideally, we have many samples of I and a and from these we can fit probability distributions. Assume that I turns out to be normally distributed with mean 1 and standard deviation 0.2, while a is uniformly distributed in the interval $[0.5, 1.5]$.

How will the uncertainty in I and a propagate through the model $u = Ie^{-at}$? That is, what is the uncertainty in u at a particular time t? This answer can easily be answered using *Monte Carlo simulation*. It means that we draw a lot of samples from the distributions for I and a. For each combination of I and a sample we compute the corresponding u value for selected values of t. Afterwards, we can for each selected t values make a histogram of all the computed u values to see what the distribution of u values look like. Figure 2.9 shows the histograms corresponding to $t = 0, 1, 3$. We see that the distribution of u values is much like a symmetric normal distribution at $t = 0$, centered around $u = 1$. At later times, the distribution gets more asymmetric and narrower. It means that the uncertainty decreases with time.

From the computed u values we can easily calculate the mean and standard deviation. The table below shows the mean and standard deviation values along with the value if we just use the formula $u = Ie^{-at}$ with the mean values of I and a: $I = 1$ and $a = 1$. As we see, there is some discrepancy between this latter (naive) computation and the mean value produced by Monte Carlo simulation.

time	mean	st.dev.	$u(t; I = a = 1)$
0	1.00	0.200	1.00
1	0.38	0.135	0.37
3	0.07	0.060	0.14

Actually, $u(t; I, a)$ becomes a stochastic variable for each t when I and a are stochastic variables, as they are in the above Monte Carlo simulation. The mean of the stochastic $u(t; I, a)$ is not equal to u with mean values of the input data, $u(t; I = a = 1)$, unless u is linear in I and a (here u is nonlinear in a).

Estimating statistical uncertainty in input data and investigating how this uncertainty propagates to uncertainty in the response of a differential equation model (or other models) are key topics in the scientific field called *uncertainty quantification*, simply known as UQ. Estimation of the statistical properties of input data can either be done directly from physical experiments, or one can find the parameter values that provide a "best fit" of model predictions with experiments. Monte Carlo simulation is a general and widely used tool to solve the associated statistical problems. The accuracy of the Monte Carlo results increases with increasing number of samples N, typically the error behaves like $N^{-1/2}$.

The computer code required to do the Monte Carlo simulation and produce the plots in Fig. 2.9 is shown below.

```
def data_errors():
    from numpy import random, mean, std
    from matplotlib.pyplot import hist
    N = 10000
    # Draw random numbers for I and a
    I_values = random.normal(1, 0.2, N)
    a_values = random.uniform(0.5, 1.5, N)
    # Compute corresponding u values for some t values
    t = [0, 1, 3]
    u_values = {}  # samples for various t values
    u_mean = {}
    u_std = {}
    for t_ in t:
        # Compute u samples corresponding to I and a samples
        u_values[t_] = [model(t_, I, a)
                        for I, a in zip(I_values, a_values)]
        u_mean[t_] = mean(u_values[t_])
        u_std[t_] = std(u_values[t_])

        figure()
        dummy1, bins, dummy2 = hist(
            u_values[t_], bins=30, range=(0, I_values.max()),
            normed=True, facecolor='green')
        #plot(bins)
        title('t=%g' % t_)
        savefig('tmp_%g.png' % t_); savefig('tmp_%g.pdf' % t_)
    # Table of mean and standard deviation values
    print 'time    mean    st.dev.'
    for t_ in t:
        print '%3g    %.2f    %.3f' % (t_, u_mean[t_], u_std[t_])
```

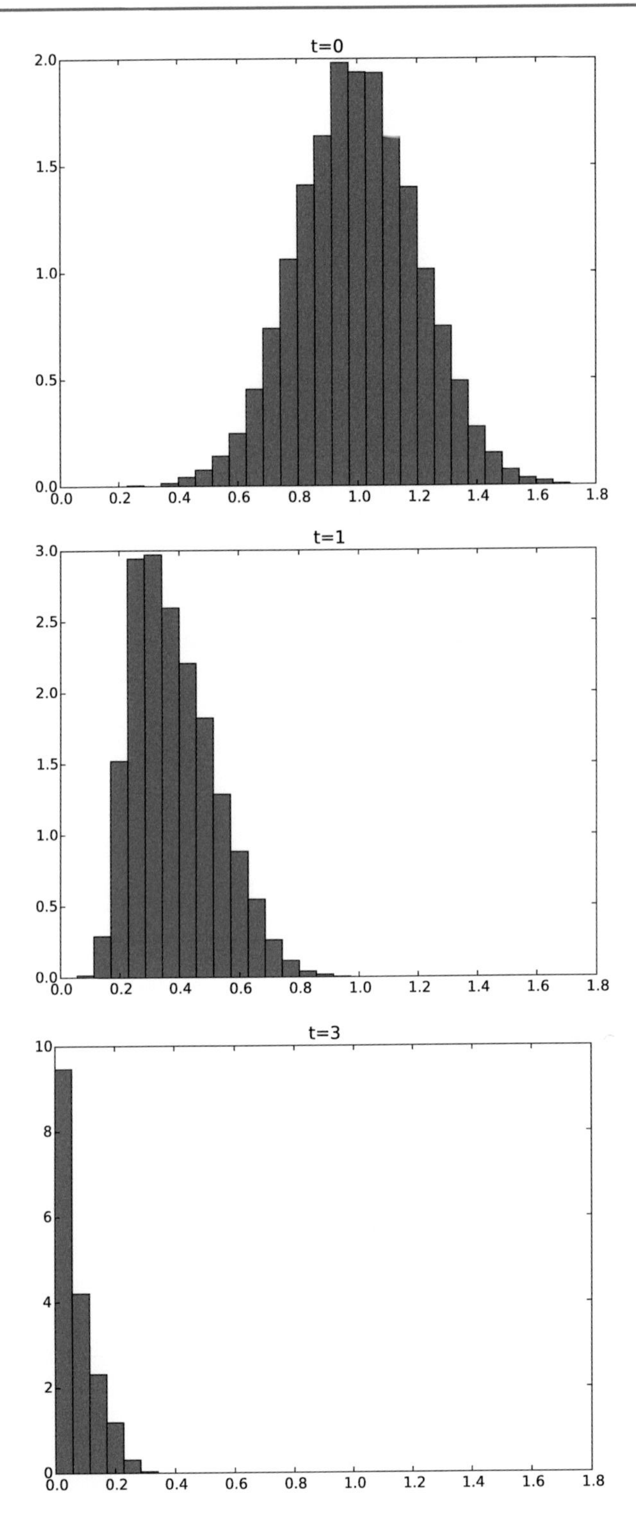

Fig. 2.9 Histogram of solution uncertainty at three time points, due to data errors

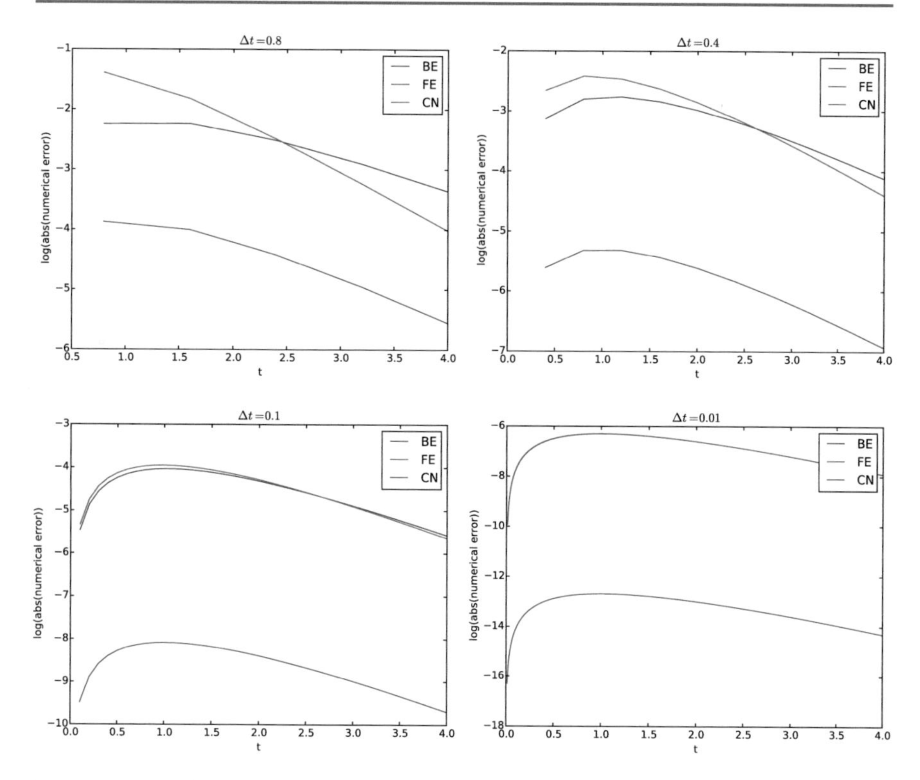

Fig. 2.10 Discretization errors in various schemes for four time step values

2.4.3 Discretization Errors

The errors implied by solving the differential equation problem by the θ-rule has been thoroughly analyzed in the previous sections. Below are some plots of the error versus time for the Forward Euler (FE), Backward Euler (BN), and Crank–Nicolson (CN) schemes for decreasing values of Δt. Since the difference in magnitude between the errors in the CN scheme versus the FE and BN schemes grows significantly as Δt is reduced (the error goes like Δt^2 for CN versus Δt for FE/BE), we have plotted the logarithm of the absolute value of the numerical error as a mesh function.

The table below presents exact figures of the discretization error for various choices of Δt and schemes.

Δt	FE	BE	CN
0.4	$9 \cdot 10^{-2}$	$6 \cdot 10^{-2}$	$5 \cdot 10^{-3}$
0.1	$2 \cdot 10^{-2}$	$2 \cdot 10^{-2}$	$3 \cdot 10^{-4}$
0.01	$2 \cdot 10^{-3}$	$2 \cdot 10^{-3}$	$3 \cdot 10^{-6}$

The computer code used to generate the plots appear next. It makes use of a `solver` function as shown in Sect. 1.2.3.

```
def discretization_errors():
    from numpy import log, abs
    I = 1
    a = 1
    T = 4
    t = linspace(0, T, 101)
    schemes = {'FE': 0, 'BE': 1, 'CN': 0.5}  # theta to scheme name
    dt_values = [0.8, 0.4, 0.1, 0.01]
    for dt in dt_values:
        figure()
        legends = []
        for scheme in schemes:
            theta = schemes[scheme]
            u, t = solver(I, a, T, dt, theta)
            u_e = model(t, I, a)
            error = u_e - u
            print '%s: dt=%.2f, %d steps, max error: %.2E' % \
                  (scheme, dt, len(u)-1, abs(error).max())
            # Plot log(error), but exclude error[0] since it is 0
            plot(t[1:], log(abs(error[1:])))
            legends.append(scheme)
        xlabel('t');  ylabel('log(abs(numerical error))')
        legend(legends, loc='upper right')
        title(r'$\Delta t=%g$' % dt)
        savefig('tmp_dt%g.png' % dt); savefig('tmp_dt%g.pdf' % dt)
```

2.4.4 Rounding Errors

Real numbers on a computer are represented by floating-point numbers[4], which means that just a finite number of digits are stored and used. Therefore, the floating-point number is an approximation to the underlying real number. When doing arithmetics with floating-point numbers, there will be small approximation errors, called round-off errors or rounding errors, that may or may not accumulate in comprehensive computations.

The cause and analysis of rounding errors are described in most books on numerical analysis, see for instance Chapter 2 in Gander et al. [1]. For very simple algorithms it is possible to theoretically establish bounds for the rounding errors, but for most algorithms one cannot know to what extent rounding errors accumulate and potentially destroy the final answer. Exercise 2.3 demonstrates the impact of rounding errors on numerical differentiation and integration.

Here is a simplest possible example of the effect of rounding errors:

```
>>> 1.0/51*51
1.0
>>> 1.0/49*49
0.9999999999999999
```

We see that the latter result is not exact, but features an error of 10^{-16}. This is the typical level of a rounding error from an arithmetic operation with the widely used 64 bit floating-point number (`float` object in Python, often called `double` or double precision in other languages). One cannot expect more accuracy than 10^{-16}. The big question is if errors at this level accumulate in a given numerical algorithm.

[4] https://en.wikipedia.org/wiki/Floating_point

What is the effect of using `float` objects and not exact arithmetics when solving differential equations? We can investigate this question through computer experiments if we have the ability to represent real numbers to a desired accuracy. Fortunately, Python has a `Decimal` object in the `decimal`[5] module that allows us to use as many digits in floating-point numbers as we like. We take 1000 digits as the true answer where rounding errors are negligible, and then we run our numerical algorithm (the Crank–Nicolson scheme to be precise) with `Decimal` objects for all real numbers and compute the maximum error arising from using 4, 16, 64, and 128 digits.

When computing with numbers around unity in size and doing $N_t = 40$ time steps, we typically get a rounding error of 10^{-d}, where d is the number of digits used. The effect of rounding errors may accumulate if we perform more operations, so increasing the number of time steps to 4000 gives a rounding error of the order 10^{-d+2}. Also, if we compute with numbers that are much larger than unity, we lose accuracy due to rounding errors. For example, for the u values implied by $I = 1000$ and $a = 100$ ($u \sim 10^3$), the rounding errors increase to about 10^{-d+3}. Below is a table summarizing a set of experiments. A rough model for the size of rounding errors is 10^{-d+q+r}, where d is the number of digits, the number of time steps is of the order 10^q time steps, and the size of the numbers in the arithmetic expressions are of order 10^r.

digits	$u \sim 1, N_t = 40$	$u \sim 1, N_t = 4000$	$u \sim 10^3, N_t = 40$	$u \sim 10^3, N_t = 4000$
4	$3.05 \cdot 10^{-4}$	$2.51 \cdot 10^{-1}$	$3.05 \cdot 10^{-1}$	$9.82 \cdot 10^{2}$
16	$1.71 \cdot 10^{-16}$	$1.42 \cdot 10^{-14}$	$1.58 \cdot 10^{-13}$	$4.84 \cdot 10^{-11}$
64	$2.99 \cdot 10^{-64}$	$1.80 \cdot 10^{-62}$	$2.06 \cdot 10^{-61}$	$1.04 \cdot 10^{-57}$
128	$1.60 \cdot 10^{-128}$	$1.56 \cdot 10^{-126}$	$2.41 \cdot 10^{-125}$	$1.07 \cdot 10^{-122}$

We realize that rounding errors are at the lowest possible level if we scale the differential equation model, see Sect. 4.1, so the numbers entering the computations are of unity in size, and if we take a small number of steps (40 steps gives a discretization error of $5 \cdot 10^{-3}$ with the Crank–Nicolson scheme). In general, rounding errors are negligible in comparison with other errors in differential equation models.

The computer code for doing the reported experiments need a new version of the `solver` function where we do arithmetics with `Decimal` objects:

```
def solver_decimal(I, a, T, dt, theta):
    """Solve u'=-a*u, u(0)=I, for t in (0,T] with steps of dt."""
    from numpy import zeros, linspace
    from decimal import Decimal as D
    dt = D(dt)
    a = D(a)
    theta = D(theta)
    Nt = int(round(D(T)/dt))
    T = Nt*dt
    u = zeros(Nt+1, dtype=object)  # array of Decimal objects
    t = linspace(0, float(T), Nt+1)

    u[0] = D(I)                     # assign initial condition
    for n in range(0, Nt):          # n=0,1,...,Nt-1
        u[n+1] = (1 - (1-theta)*a*dt)/(1 + theta*dt*a)*u[n]
    return u, t
```

[5] https://docs.python.org/2/library/decimal.html

The function below carries out the experiments. We can conveniently set the number of digits as we want through the `decimal.getcontext().prec` variable.

```
def rounding_errors(I=1, a=1, T=4, dt=0.1):
    import decimal
    from numpy import log, array, abs
    digits_values = [4, 16, 64, 128]
    # "Exact" arithmetics is taken as 1000 decimals here
    decimal.getcontext().prec = 1000
    u_e, t = solver_decimal(I=I, a=a, T=T, dt=dt, theta=0.5)
    for digits in digits_values:
        decimal.getcontext().prec = digits  # set no of digits
        u, t = solver_decimal(I=I, a=a, T=T, dt=dt, theta=0.5)
        error = u_e - u
        error = array(error[1:], dtype=float)
        print '%d digits, %d steps, max abs(error): %.2E' % \
              (digits, len(u)-1, abs(error).max())
```

2.4.5 Discussion of the Size of Various Errors

The previous computational examples of model, data, discretization, and rounding errors are tied to one particular mathematical problem, so it is in principle dangerous to make general conclusions. However, the illustrations made point to some common trends that apply to differential equation models.

First, rounding errors have very little impact compared to the other types of errors. Second, numerical errors are in general smaller than model and data errors, but more importantly, numerical errors are often well understood and can be reduced by just increasing the computational work (in our example by taking more smaller time steps).

Third, data errors may be significant, and it also takes a significant amount of computational work to quantify them and their impact on the solution. Many types of input data are also difficult or impossible to measure, so finding suitable values requires tuning of the data and the model to a known (measured) response. Nevertheless, even if the predictive precision of a model is limited because of severe errors or uncertainty in input data, the model can still be of high value for investigating qualitative properties of the underlying phenomenon. Through computer experiments with synthetic input data one can understand a lot of the science or engineering that goes into the model.

Fourth, model errors are the most challenging type of error to deal with. Simplicity of model is in general preferred over complexity, but adding complexity is often the only way to improve the predictive capabilities of a model. More complexity usually also means a need for more input data and consequently the danger of increasing data errors.

2.5 Exercises

Problem 2.1: Visualize the accuracy of finite differences

The purpose of this exercise is to visualize the accuracy of finite difference approximations of the derivative of a given function. For any finite difference approximation, take the Forward Euler difference as an example, and any specific function, take $u = e^{-at}$, we may introduce an error fraction

$$E = \frac{[D_t^+ u]^n}{u'(t_n)} = \frac{\exp(-a(t_n + \Delta t)) - \exp(-at_n)}{-a \exp(-at_n)\Delta t}$$
$$= \frac{1}{a\Delta t}(1 - \exp(-a\Delta t)),$$

and view E as a function of Δt. We expect that $\lim_{\Delta t \to 0} E = 1$, while E may deviate significantly from unity for large Δt. How the error depends on Δt is best visualized in a graph where we use a logarithmic scale for Δt, so we can cover many orders of magnitude of that quantity. Here is a code segment creating an array of 100 intervals, on the logarithmic scale, ranging from 10^{-6} to $10^{-0.5}$ and then plotting E versus $p = a\Delta t$ with logarithmic scale on the p axis:

```
from numpy import logspace, exp
from matplotlib.pyplot import semilogx
p = logspace(-6, -0.5, 101)
y = (1-exp(-p))/p
semilogx(p, y)
```

Illustrate such errors for the finite difference operators $[D_t^+ u]^n$ (forward), $[D_t^- u]^n$ (backward), and $[D_t u]^n$ (centered) in the same plot.

Perform a Taylor series expansions of the error fractions and find the leading order r in the expressions of type $1 + Cp^r + \mathcal{O}(p^{r+1})$, where C is some constant.

Hint To save manual calculations and learn more about symbolic computing, make functions for the three difference operators and use `sympy` to perform the symbolic differences, differentiation, and Taylor series expansion. To plot a symbolic expression `E` against `p`, convert the expression to a Python function first: `E = sympy.lamdify([p], E)`.

Filename: `decay_plot_fd_error`.

Problem 2.2: Explore the θ-rule for exponential growth

This exercise asks you to solve the ODE $u' = -au$ with $a < 0$ such that the ODE models exponential growth instead of exponential decay. A central theme is to investigate numerical artifacts and non-physical solution behavior.

a) Set $a = -1$ and run experiments with $\theta = 0, 0.5, 1$ for various values of Δt to uncover numerical artifacts. Recall that the exact solution is a monotone, growing function when $a < 0$. Oscillations or significantly wrong growth are signs of wrong qualitative behavior.
From the experiments, select four values of Δt that demonstrate the kind of numerical solutions that are characteristic for this model.
b) Write up the amplification factor and plot it for $\theta = 0, 0.5, 1$ together with the exact one for $a\Delta t < 0$. Use the plot to explain the observations made in the experiments.

Hint Modify the `decay_ampf_plot.py`[6] code (in the `src/analysis` directory). Filename: `exponential_growth`.

Problem 2.3: Explore rounding errors in numerical calculus

a) Compute the absolute values of the errors in the numerical derivative of e^{-t} at $t = \frac{1}{2}$ for three types of finite difference approximations: a forward difference, a backward difference, and a centered difference, for $\Delta t = 2^{-k}$, $k = 0, 4, 8, 12, \ldots, 60$. When do rounding errors destroy the accuracy?
b) Compute the absolute values of the errors in the numerical approximation of $\int_0^4 e^{-t}dt$ using the Trapezoidal and the Midpoint integration methods. Make a table of the errors for $n = 2^k$ intervals, $k = 1, 3, 5, \ldots, 21$. Is there any impact of rounding errors?

Hint The Trapezoidal rule for $\int_a^b f(x)dx$ reads

$$\int\limits_a^b f(x)dx \approx h\left(\frac{1}{2}f(a) + \frac{1}{2}f(b) + \sum_{i=1}^{n-1} f(a+ih)\right), \quad h = \frac{b-a}{n}.$$

The Midpoint rule is

$$\int\limits_a^b f(x)dx \approx h\sum_{i=1}^{n} f\left(a + \left(i + \frac{1}{2}\right)h\right).$$

Filename: `rounding`.

[6] http://tinyurl.com/ofkw6kc/analysis/decay_ampf_plot.py

Generalizations

3

It is time to consider generalizations of the simple decay model $u' = -au$ and also to look at additional numerical solution methods. We consider first variable coefficients, $u' = a(t)u + b(t)$, and later a completely general scalar ODE $u' = f(u,t)$ and its generalization to a system of such general ODEs. Among numerical methods, we treat implicit multi-step methods, and several families of explicit methods: Leapfrog schemes, Runge–Kutta methods, and Adams–Bashforth formulas.

3.1 Model Extensions

This section looks at the generalizations to $u' = -a(t)u$ and $u' = -a(t)u + b(t)$. We sketch the corresponding implementations of the θ-rule for such variable-coefficient ODEs. Verification can no longer make use of an exact solution of the numerical problem so we make use of manufactured solutions, for deriving an exact solution of the ODE problem, and then we can compute empirical convergence rates for the method and see if these coincide with the expected rates from theory. Finally, we see how our numerical methods can be applied to systems of ODEs.

The example programs associated with this chapter are found in the directory `src/genz`[1].

3.1.1 Generalization: Including a Variable Coefficient

In the ODE for decay, $u' = -au$, we now consider the case where a depends on time:

$$u'(t) = -a(t)u(t), \quad t \in (0,T], \quad u(0) = I\,. \tag{3.1}$$

A Forward Euler scheme consists of evaluating (3.1) at $t = t_n$ and approximating the derivative with a forward difference $[D_t^+ u]^n$:

$$\frac{u^{n+1} - u^n}{\Delta t} = -a(t_n)u^n\,. \tag{3.2}$$

[1] http://tinyurl.com/ofkw6kc/genz

H.P. Langtangen, *Finite Difference Computing with Exponential Decay Models*, Lecture Notes in Computational Science and Engineering 110,
DOI 10.1007/978-3-319-29439-1_3

The Backward Euler scheme becomes

$$\frac{u^n - u^{n-1}}{\Delta t} = -a(t_n)u^n \,. \tag{3.3}$$

The Crank–Nicolson method builds on sampling the ODE at $t_{n+\frac{1}{2}}$. We can evaluate a at $t_{n+\frac{1}{2}}$ and use an average for u at times t_n and t_{n+1}:

$$\frac{u^{n+1} - u^n}{\Delta t} = -a(t_{n+\frac{1}{2}})\frac{1}{2}(u^n + u^{n+1}) \,. \tag{3.4}$$

Alternatively, we can use an average for the product au:

$$\frac{u^{n+1} - u^n}{\Delta t} = -\frac{1}{2}(a(t_n)u^n + a(t_{n+1})u^{n+1}) \,. \tag{3.5}$$

The θ-rule unifies the three mentioned schemes. One version is to have a evaluated at the weighted time point $(1-\theta)t_n + \theta t_{n+1}$,

$$\frac{u^{n+1} - u^n}{\Delta t} = -a((1-\theta)t_n + \theta t_{n+1})((1-\theta)u^n + \theta u^{n+1}) \,. \tag{3.6}$$

Another possibility is to apply a weighted average for the product au,

$$\frac{u^{n+1} - u^n}{\Delta t} = -(1-\theta)a(t_n)u^n - \theta a(t_{n+1})u^{n+1} \,. \tag{3.7}$$

With the finite difference operator notation the Forward Euler and Backward Euler schemes can be summarized as

$$[D_t^+ u = -au]^n, \tag{3.8}$$

$$[D_t^- u = -au]^n \,. \tag{3.9}$$

The Crank–Nicolson and θ schemes depend on whether we evaluate a at the sample point for the ODE or if we use an average. The various versions are written as

$$[D_t u = -a\overline{u}^t]^{n+\frac{1}{2}}, \tag{3.10}$$

$$[D_t u = -\overline{au}^t]^{n+\frac{1}{2}}, \tag{3.11}$$

$$[D_t u = -a\overline{u}^{t,\theta}]^{n+\theta}, \tag{3.12}$$

$$[D_t u = -\overline{au}^{t,\theta}]^{n+\theta} \,. \tag{3.13}$$

3.1.2 Generalization: Including a Source Term

A further extension of the model ODE is to include a source term $b(t)$:

$$u'(t) = -a(t)u(t) + b(t), \quad t \in (0,T], \quad u(0) = I \,. \tag{3.14}$$

The time point where we sample the ODE determines where $b(t)$ is evaluated. For the Crank–Nicolson scheme and the θ-rule we have a choice of whether to evaluate $a(t)$ and $b(t)$ at the correct point or use an average. The chosen strategy becomes particularly clear if we write up the schemes in the operator notation:

$$[D_t^+ u = -au + b]^n, \tag{3.15}$$

$$[D_t^- u = -au + b]^n, \tag{3.16}$$

$$[D_t u = -a\overline{u}^t + b]^{n+\frac{1}{2}}, \tag{3.17}$$

$$[D_t u = \overline{-au + b}^t]^{n+\frac{1}{2}}, \tag{3.18}$$

$$[D_t u = -a\overline{u}^{t,\theta} + b]^{n+\theta}, \tag{3.19}$$

$$[D_t u = \overline{-au + b}^{t,\theta}]^{n+\theta}. \tag{3.20}$$

3.1.3 Implementation of the Generalized Model Problem

Deriving the θ-rule formula Writing out the θ-rule in (3.20), using (1.44) and (1.45), we get

$$\frac{u^{n+1} - u^n}{\Delta t} = \theta(-a^{n+1}u^{n+1} + b^{n+1})) + (1-\theta)(-a^n u^n + b^n)), \tag{3.21}$$

where a^n means evaluating a at $t = t_n$ and similar for a^{n+1}, b^n, and b^{n+1}. We solve for u^{n+1}:

$$u^{n+1} = ((1 - \Delta t(1-\theta)a^n)u^n + \Delta t(\theta b^{n+1} + (1-\theta)b^n))(1 + \Delta t\theta a^{n+1})^{-1}. \tag{3.22}$$

Python code Here is a suitable implementation of (3.21) where $a(t)$ and $b(t)$ are given as Python functions:

```
def solver(I, a, b, T, dt, theta):
    """
    Solve u'=-a(t)*u + b(t), u(0)=I,
    for t in (0,T] with steps of dt.
    a and b are Python functions of t.
    """
    dt = float(dt)            # avoid integer division
    Nt = int(round(T/dt))     # no of time intervals
    T = Nt*dt                 # adjust T to fit time step dt
    u = zeros(Nt+1)           # array of u[n] values
    t = linspace(0, T, Nt+1)  # time mesh

    u[0] = I                  # assign initial condition
    for n in range(0, Nt):    # n=0,1,...,Nt-1
        u[n+1] = ((1 - dt*(1-theta)*a(t[n]))*u[n] + \
                  dt*(theta*b(t[n+1]) + (1-theta)*b(t[n])))/\
                  (1 + dt*theta*a(t[n+1]))
    return u, t
```

This function is found in the file `decay_vc.py`[2] (`vc` stands for “variable coefficients”).

[2] http://tinyurl.com/ofkw6kc/genz/decay_vc.py

Coding of variable coefficients The `solver` function shown above demands the arguments `a` and `b` to be Python functions of time `t`, say

```
def a(t):
    return a_0 if t < tp else k*a_0

def b(t):
    return 1
```

Here, `a(t)` has three parameters `a0`, `tp`, and `k`, which must be global variables.

A better implementation, which avoids global variables, is to represent `a` by a class where the parameters are attributes and where a *special method* `__call__` evaluates $a(t)$:

```
class A:
    def __init__(self, a0=1, k=2):
        self.a0, self.k = a0, k

    def __call__(self, t):
        return self.a0 if t < self.tp else self.k*self.a0

a = A(a0=2, k=1)  # a behaves as a function a(t)
```

For quick tests it is cumbersome to write a complete function or a class. The *lambda function* construction in Python is then convenient. For example,

```
a = lambda t: a_0 if t < tp else k*a_0
```

is equivalent to the `def a(t)` definition above. In general,

```
f = lambda arg1, arg2, ...: expression
```

is equivalent to

```
def f(arg1, arg2, ...):
    return expression
```

One can use lambda functions directly in calls. Say we want to solve $u' = -u + 1$, $u(0) = 2$:

```
u, t = solver(2, lambda t: 1, lambda t: 1, T, dt, theta)
```

Whether to use a plain function, a class, or a lambda function depends on the programmer's taste. Lazy programmers prefer the lambda construct, while very safe programmers go for the class solution.

3.1.4 Verifying a Constant Solution

An extremely useful partial verification method is to construct a test problem with a very simple solution, usually $u =$ const. Especially the initial debugging of a program code can benefit greatly from such tests, because 1) all relevant numerical methods will exactly reproduce a constant solution, 2) many of the intermediate calculations are easy to control by hand for a constant u, and 3) even a constant u can uncover many bugs in an implementation.

The only constant solution for the problem $u' = -au$ is $u = 0$, but too many bugs can escape from that trivial solution. It is much better to search for a problem where $u = C = \text{const} \neq 0$. Then $u' = -a(t)u + b(t)$ is more appropriate: with $u = C$ we can choose any $a(t)$ and set $b = a(t)C$ and $I = C$. An appropriate test function is

```
def test_constant_solution():
    """
    Test problem where u=u_const is the exact solution, to be
    reproduced (to machine precision) by any relevant method.
    """
    def u_exact(t):
        return u_const

    def a(t):
        return 2.5*(1+t**3)  # can be arbitrary

    def b(t):
        return a(t)*u_const

    u_const = 2.15
    theta = 0.4; I = u_const; dt = 4
    Nt = 4  # enough with a few steps
    u, t = solver(I=I, a=a, b=b, T=Nt*dt, dt=dt, theta=theta)
    print u
    u_e = u_exact(t)
    difference = abs(u_e - u).max()  # max deviation
    tol = 1E-14
    assert difference < tol
```

An interesting question is what type of bugs that will make the computed u^n deviate from the exact solution C. Fortunately, the updating formula and the initial condition must be absolutely correct for the test to pass! Any attempt to make a wrong indexing in terms like `a(t[n])` or any attempt to introduce an erroneous factor in the formula creates a solution that is different from C.

3.1.5 Verification via Manufactured Solutions

Following the idea of the previous section, we can choose any formula as the exact solution, insert the formula in the ODE problem and fit the data $a(t)$, $b(t)$, and I to make the chosen formula fulfill the equation. This powerful technique for generating exact solutions is very useful for verification purposes and known as the *method of manufactured solutions*, often abbreviated MMS.

One common choice of solution is a linear function in the independent variable(s). The rationale behind such a simple variation is that almost any relevant numerical solution method for differential equation problems is able to reproduce a linear function exactly to machine precision (if u is about unity in size; precision is lost if u takes on large values, see Exercise 3.1). The linear solution also makes some stronger demands to the numerical method and the implementation than the constant solution used in Sect. 3.1.4, at least in more complicated applications. Still, the constant solution is often ideal for initial debugging before proceeding with a linear solution.

We choose a linear solution $u(t) = ct + d$. From the initial condition it follows that $d = I$. Inserting this u in the left-hand side of (3.14), i.e., the ODE, we get

$$c = -a(t)u + b(t)\,.$$

Any function $u = ct + I$ is then a correct solution if we choose

$$b(t) = c + a(t)(ct + I)\,.$$

With this $b(t)$ there are no restrictions on $a(t)$ and c.

Let us prove that such a linear solution obeys the numerical schemes. To this end, we must check that $u^n = ca(t_n)(ct_n + I)$ fulfills the discrete equations. For these calculations, and later calculations involving linear solutions inserted in finite difference schemes, it is convenient to compute the action of a difference operator on a linear function t:

$$[D_t^+ t]^n = \frac{t_{n+1} - t_n}{\Delta t} = 1, \tag{3.23}$$

$$[D_t^- t]^n = \frac{t_n - t_{n-1}}{\Delta t} = 1, \tag{3.24}$$

$$[D_t t]^n = \frac{t_{n+\frac{1}{2}} - t_{n-\frac{1}{2}}}{\Delta t} = \frac{(n + \frac{1}{2})\Delta t - (n - \frac{1}{2})\Delta t}{\Delta t} = 1\,. \tag{3.25}$$

Clearly, all three finite difference approximations to the derivative are exact for $u(t) = t$ or its mesh function counterpart $u^n = t_n$.

The difference equation for the Forward Euler scheme

$$[D_t^+ u = -au + b]^n,$$

with $a^n = a(t_n)$, $b^n = c + a(t_n)(ct_n + I)$, and $u^n = ct_n + I$ then results in

$$c = -a(t_n)(ct_n + I) + c + a(t_n)(ct_n + I) = c$$

which is always fulfilled. Similar calculations can be done for the Backward Euler and Crank–Nicolson schemes, or the θ-rule for that matter. In all cases, $u^n = ct_n + I$ is an exact solution of the discrete equations. That is why we should expect that $u^n - u_e(t_n) = 0$ mathematically and $|u^n - u_e(t_n)|$ less than a small number about the machine precision for $n = 0, \ldots, N_t$.

The following function offers an implementation of this verification test based on a linear exact solution:

```
def test_linear_solution():
    """
    Test problem where u=c*t+I is the exact solution, to be
    reproduced (to machine precision) by any relevant method.
    """
```

```
def u_exact(t):
    return c*t + I

def a(t):
    return t**0.5  # can be arbitrary

def b(t):
    return c + a(t)*u_exact(t)

theta = 0.4; I = 0.1; dt = 0.1; c = -0.5
T = 4
Nt = int(T/dt)  # no of steps
u, t = solver(I=I, a=a, b=b, T=Nt*dt, dt=dt, theta=theta)
u_e = u_exact(t)
difference = abs(u_e - u).max()  # max deviation
print difference
tol = 1E-14  # depends on c!
assert difference < tol
```

Any error in the updating formula makes this test fail!

Choosing more complicated formulas as the exact solution, say $\cos(t)$, will not make the numerical and exact solution coincide to machine precision, because finite differencing of $\cos(t)$ does not exactly yield the exact derivative $-\sin(t)$. In such cases, the verification procedure must be based on measuring the convergence rates as exemplified in Sect. 3.1.6. Convergence rates can be computed as long as one has an exact solution of a problem that the solver can be tested on, but this can always be obtained by the method of manufactured solutions.

3.1.6 Computing Convergence Rates

We expect that the error E in the numerical solution is reduced if the mesh size Δt is decreased. More specifically, many numerical methods obey a power-law relation between E and Δt:

$$E = C\Delta t^r, \tag{3.26}$$

where C and r are (usually unknown) constants independent of Δt. The formula (3.26) is viewed as an asymptotic model valid for sufficiently small Δt. How small is normally hard to estimate without doing numerical estimations of r.

The parameter r is known as the *convergence rate*. For example, if the convergence rate is 2, halving Δt reduces the error by a factor of 4. Diminishing Δt then has a greater impact on the error compared with methods that have $r = 1$. For a given value of r, we refer to the method as of r-th order. First- and second-order methods are most common in scientific computing.

Estimating r There are two alternative ways of estimating C and r based on a set of m simulations with corresponding pairs $(\Delta t_i, E_i)$, $i = 0, \ldots, m-1$, and $\Delta t_i < \Delta t_{i-1}$ (i.e., decreasing cell size).

1. Take the logarithm of (3.26), $\ln E = r \ln \Delta t + \ln C$, and fit a straight line to the data points $(\Delta t_i, E_i)$, $i = 0, \ldots, m-1$.

2. Consider two consecutive experiments, $(\Delta t_i, E_i)$ and $(\Delta t_{i-1}, E_{i-1})$. Dividing the equation $E_{i-1} = C\Delta t_{i-1}^r$ by $E_i = C\Delta t_i^r$ and solving for r yields

$$r_{i-1} = \frac{\ln(E_{i-1}/E_i)}{\ln(\Delta t_{i-1}/\Delta t_i)} \tag{3.27}$$

for $i = 1, \ldots, m-1$. Note that we have introduced a subindex $i-1$ on r in (3.27) because r estimated from a pair of experiments must be expected to change with i.

The disadvantage of method 1 is that (3.26) might not be valid for the coarsest meshes (largest Δt values). Fitting a line to all the data points is then misleading. Method 2 computes convergence rates for pairs of experiments and allows us to see if the sequence r_i converges to some value as $i \to m-2$. The final r_{m-2} can then be taken as the convergence rate. If the coarsest meshes have a differing rate, the corresponding time steps are probably too large for (3.26) to be valid. That is, those time steps lie outside the asymptotic range of Δt values where the error behaves like (3.26).

Implementation We can compute $r_0, r_1, \ldots, r_{m-2}$ from E_i and Δt_i by the following function

```
def compute_rates(dt_values, E_values):
    m = len(dt_values)
    r = [log(E_values[i-1]/E_values[i])/
         log(dt_values[i-1]/dt_values[i])
         for i in range(1, m, 1)]
    # Round to two decimals
    r = [round(r_, 2) for r_ in r]
    return r
```

Experiments with a series of time step values and $\theta = 0, 1, 0.5$ can be set up as follows, here embedded in a real test function:

```
def test_convergence_rates():
    # Create a manufactured solution
    # define u_exact(t), a(t), b(t)

    dt_values = [0.1*2**(-i) for i in range(7)]
    I = u_exact(0)

    for theta in (0, 1, 0.5):
        E_values = []
        for dt in dt_values:
            u, t = solver(I=I, a=a, b=b, T=6, dt=dt, theta=theta)
            u_e = u_exact(t)
            e = u_e - u
            E = sqrt(dt*sum(e**2))
            E_values.append(E)
        r = compute_rates(dt_values, E_values)
        print 'theta=%g, r: %s' % (theta, r)
        expected_rate = 2 if theta == 0.5 else 1
        tol = 0.1
        diff = abs(expected_rate - r[-1])
        assert diff < tol
```

The manufactured solution is conveniently computed by sympy. Let us choose $u_e(t) = \sin(t)e^{-2t}$ and $a(t) = t^2$. This implies that we must fit b as $b(t) = u'(t) - a(t)$. We first compute with sympy expressions and then we convert the exact solution, a, and b to Python functions that we can use in the subsequent numerical computing:

```
# Create a manufactured solution with sympy
import sympy as sym
t = sym.symbols('t')
u_e = sym.sin(t)*sym.exp(-2*t)
a = t**2
b = sym.diff(u_e, t) + a*u_exact

# Turn sympy expressions into Python function
u_exact = sym.lambdify([t], u_e, modules='numpy')
a = sym.lambdify([t], a, modules='numpy')
b = sym.lambdify([t], b, modules='numpy')
```

The complete code is found in the function test_convergence_rates in the file decay_vc.py[3].

Running this code gives the output

Terminal

```
theta=0, r: [1.06, 1.03, 1.01, 1.01, 1.0, 1.0]
theta=1, r: [0.94, 0.97, 0.99, 0.99, 1.0, 1.0]
theta=0.5, r: [2.0, 2.0, 2.0, 2.0, 2.0, 2.0]
```

We clearly see how the convergence rates approach the expected values.

Why convergence rates are important

The strong practical application of computing convergence rates is for verification: wrong convergence rates point to errors in the code, and correct convergence rates bring strong support for a correct implementation. Experience shows that bugs in the code easily destroy the expected convergence rate.

3.1.7 Extension to Systems of ODEs

Many ODE models involve more than one unknown function and more than one equation. Here is an example of two unknown functions $u(t)$ and $v(t)$:

$$u' = au + bv, \tag{3.28}$$

$$v' = cu + dv, \tag{3.29}$$

for constants a, b, c, d. Applying the Forward Euler method to each equation results in a simple updating formula:

$$u^{n+1} = u^n + \Delta t(au^n + bv^n), \tag{3.30}$$

$$v^{n+1} = u^n + \Delta t(cu^n + dv^n). \tag{3.31}$$

[3] http://tinyurl.com/ofkw6kc/genz/decay_vc.py

On the other hand, the Crank–Nicolson or Backward Euler schemes result in a 2×2 linear system for the new unknowns. The latter scheme becomes

$$u^{n+1} = u^n + \Delta t(au^{n+1} + bv^{n+1}), \tag{3.32}$$

$$v^{n+1} = v^n + \Delta t(cu^{n+1} + dv^{n+1})\,. \tag{3.33}$$

Collecting u^{n+1} as well as v^{n+1} on the left-hand side results in

$$(1 - \Delta ta)u^{n+1} + bv^{n+1} = u^n, \tag{3.34}$$

$$cu^{n+1} + (1 - \Delta td)v^{n+1} = v^n, \tag{3.35}$$

which is a system of two coupled, linear, algebraic equations in two unknowns. These equations can be solved by hand (using standard techniques for two algebraic equations with two unknowns x and y), resulting in explicit formulas for u^{n+1} and v^{n+1} that can be directly implemented. For systems of ODEs with many equations and unknowns, one will express the coupled equations at each time level in matrix form and call software for numerical solution of linear systems of equations.

3.2 General First-Order ODEs

We now turn the attention to general, nonlinear ODEs and systems of such ODEs. Our focus is on numerical methods that can be readily reused for time-discretization of PDEs, and diffusion PDEs in particular. The methods are just briefly listed, and we refer to the rich literature for more detailed descriptions and analysis – the books [2–4, 12] are all excellent resources on numerical methods for ODEs. We also demonstrate the Odespy Python interface to a range of different software for general first-order ODE systems.

3.2.1 Generic Form of First-Order ODEs

ODEs are commonly written in the generic form

$$u' = f(u,t), \quad u(0) = I, \tag{3.36}$$

where $f(u,t)$ is some prescribed function. As an example, our most general exponential decay model (3.14) has $f(u,t) = -a(t)u(t) + b(t)$.

The unknown u in (3.36) may either be a scalar function of time t, or a vector valued function of t in case of a *system of ODEs* with m unknown components:

$$u(t) = (u^{(0)}(t), u^{(1)}(t), \ldots, u^{(m-1)}(t))\,.$$

In that case, the right-hand side is a vector-valued function with m components,

$$\begin{aligned} f(u,t) = (&f^{(0)}(u^{(0)}(t), \ldots, u^{(m-1)}(t)),\\ &f^{(1)}(u^{(0)}(t), \ldots, u^{(m-1)}(t)),\\ &\vdots\\ &f^{(m-1)}(u^{(0)}(t), \ldots, u^{(m-1)}(t)))\,. \end{aligned}$$

Actually, any system of ODEs can be written in the form (3.36), but higher-order ODEs then need auxiliary unknown functions to enable conversion to a first-order system.

Next we list some well-known methods for $u' = f(u,t)$, valid both for a single ODE (scalar u) and systems of ODEs (vector u).

3.2.2 The θ-Rule

The θ-rule scheme applied to $u' = f(u,t)$ becomes

$$\frac{u^{n+1} - u^n}{\Delta t} = \theta f(u^{n+1}, t_{n+1}) + (1-\theta) f(u^n, t_n)\,. \tag{3.37}$$

Bringing the unknown u^{n+1} to the left-hand side and the known terms on the right-hand side gives

$$u^{n+1} - \Delta t \theta f(u^{n+1}, t_{n+1}) = u^n + \Delta t (1-\theta) f(u^n, t_n)\,. \tag{3.38}$$

For a general f (not linear in u), this equation is *nonlinear* in the unknown u^{n+1} unless $\theta = 0$. For a scalar ODE ($m = 1$), we have to solve a single nonlinear algebraic equation for u^{n+1}, while for a system of ODEs, we get a system of coupled, nonlinear algebraic equations. Newton's method is a popular solution approach in both cases. Note that with the Forward Euler scheme ($\theta = 0$) we do not have to deal with nonlinear equations, because in that case we have an explicit updating formula for u^{n+1}. This is known as an *explicit* scheme. With $\theta \neq 1$ we have to solve (systems of) algebraic equations, and the scheme is said to be *implicit*.

3.2.3 An Implicit 2-Step Backward Scheme

The implicit backward method with 2 steps applies a three-level backward difference as approximation to $u'(t)$,

$$u'(t_{n+1}) \approx \frac{3u^{n+1} - 4u^n + u^{n-1}}{2\Delta t},$$

which is an approximation of order Δt^2 to the first derivative. The resulting scheme for $u' = f(u,t)$ reads

$$u^{n+1} = \frac{4}{3}u^n - \frac{1}{3}u^{n-1} + \frac{2}{3}\Delta t f(u^{n+1}, t_{n+1})\,. \tag{3.39}$$

Higher-order versions of the scheme (3.39) can be constructed by including more time levels. These schemes are known as the Backward Differentiation Formulas (BDF), and the particular version (3.39) is often referred to as BDF2.

Note that the scheme (3.39) is implicit and requires solution of nonlinear equations when f is nonlinear in u. The standard 1st-order Backward Euler method or the Crank–Nicolson scheme can be used for the first step.

3.2.4 Leapfrog Schemes

The ordinary Leapfrog scheme The derivative of u at some point t_n can be approximated by a central difference over two time steps,

$$u'(t_n) \approx \frac{u^{n+1} - u^{n-1}}{2\Delta t} = [D_{2t}u]^n \tag{3.40}$$

which is an approximation of second order in Δt. The scheme can then be written as

$$[D_{2t}u = f(u,t)]^n,$$

in operator notation. Solving for u^{n+1} gives

$$u^{n+1} = u^{n-1} + 2\Delta t f(u^n, t_n)\,. \tag{3.41}$$

Observe that (3.41) is an explicit scheme, and that a nonlinear f (in u) is trivial to handle since it only involves the known u^n value. Some other scheme must be used as starter to compute u^1, preferably the Forward Euler scheme since it is also explicit.

The filtered Leapfrog scheme Unfortunately, the Leapfrog scheme (3.41) will develop growing oscillations with time (see Problem 3.6). A remedy for such undesired oscillations is to introduce a *filtering technique*. First, a standard Leapfrog step is taken, according to (3.41), and then the previous u^n value is adjusted according to

$$u^n \leftarrow u^n + \gamma(u^{n-1} - 2u^n + u^{n+1})\,. \tag{3.42}$$

The γ-terms will effectively damp oscillations in the solution, especially those with short wavelength (like point-to-point oscillations). A common choice of γ is 0.6 (a value used in the famous NCAR Climate Model).

3.2.5 The 2nd-Order Runge–Kutta Method

The two-step scheme

$$u^* = u^n + \Delta t f(u^n, t_n), \tag{3.43}$$

$$u^{n+1} = u^n + \Delta t \frac{1}{2}\left(f(u^n, t_n) + f(u^*, t_{n+1})\right), \tag{3.44}$$

essentially applies a Crank–Nicolson method (3.44) to the ODE, but replaces the term $f(u^{n+1}, t_{n+1})$ by a prediction $f(u^*, t_{n+1})$ based on a Forward Euler step (3.43). The scheme (3.43)–(3.44) is known as Huen's method, but is also a 2nd-order Runge–Kutta method. The scheme is explicit, and the error is expected to behave as Δt^2.

3.2.6 A 2nd-Order Taylor-Series Method

One way to compute u^{n+1} given u^n is to use a Taylor polynomial. We may write up a polynomial of 2nd degree:

$$u^{n+1} = u^n + u'(t_n)\Delta t + \frac{1}{2}u''(t_n)\Delta t^2 .$$

From the equation $u' = f(u,t)$ it follows that the derivatives of u can be expressed in terms of f and its derivatives:

$$\begin{aligned}
u'(t_n) &= f(u^n, t_n),\\
u''(t_n) &= \frac{\partial f}{\partial u}(u^n, t_n)u'(t_n) + \frac{\partial f}{\partial t}\\
&= f(u^n, t_n)\frac{\partial f}{\partial u}(u^n, t_n) + \frac{\partial f}{\partial t},
\end{aligned}$$

resulting in the scheme

$$u^{n+1} = u^n + f(u^n, t_n)\Delta t + \frac{1}{2}\left(f(u^n, t_n)\frac{\partial f}{\partial u}(u^n, t_n) + \frac{\partial f}{\partial t}\right)\Delta t^2 . \tag{3.45}$$

More terms in the series could be included in the Taylor polynomial to obtain methods of higher order than 2.

3.2.7 The 2nd- and 3rd-Order Adams–Bashforth Schemes

The following method is known as the 2nd-order Adams–Bashforth scheme:

$$u^{n+1} = u^n + \frac{1}{2}\Delta t\left(3f(u^n, t_n) - f(u^{n-1}, t_{n-1})\right) . \tag{3.46}$$

The scheme is explicit and requires another one-step scheme to compute u^1 (the Forward Euler scheme or Heun's method, for instance). As the name implies, the error behaves like Δt^2.

Another explicit scheme, involving four time levels, is the 3rd-order Adams–Bashforth scheme

$$u^{n+1} = u^n + \frac{1}{12}\left(23f(u^n, t_n) - 16f(u^{n-1}, t_{n-1}) + 5f(u^{n-2}, t_{n-2})\right) . \tag{3.47}$$

The numerical error is of order Δt^3, and the scheme needs some method for computing u^1 and u^2.

More general, higher-order Adams–Bashforth schemes (also called *explicit Adams methods*) compute u^{n+1} as a linear combination of f at $k+1$ previous time steps:

$$u^{n+1} = u^n + \sum_{j=0}^{k}\beta_j f(u^{n-j}, t_{n-j}),$$

where β_j are known coefficients.

3.2.8 The 4th-Order Runge–Kutta Method

The perhaps most widely used method to solve ODEs is the 4th-order Runge–Kutta method, often called RK4. Its derivation is a nice illustration of common numerical approximation strategies, so let us go through the steps in detail to learn about algorithmic development.

The starting point is to integrate the ODE $u' = f(u,t)$ from t_n to t_{n+1}:

$$u(t_{n+1}) - u(t_n) = \int_{t_n}^{t_{n+1}} f(u(t),t)dt\,.$$

We want to compute $u(t_{n+1})$ and regard $u(t_n)$ as known. The task is to find good approximations for the integral, since the integrand involves the unknown u between t_n and t_{n+1}.

The integral can be approximated by the famous Simpson's rule[4]:

$$\int_{t_n}^{t_{n+1}} f(u(t),t)dt \approx \frac{\Delta t}{6}\left(f^n + 4f^{n+\frac{1}{2}} + f^{n+1}\right).$$

The problem now is that we do not know $f^{n+\frac{1}{2}} = f(u^{n+\frac{1}{2}}, t_{n+\frac{1}{2}})$ and $f^{n+1} = (u^{n+1}, t_{n+1})$ as we know only u^n and hence f^n. The idea is to use various approximations for $f^{n+\frac{1}{2}}$ and f^{n+1} based on well-known schemes for the ODE in the intervals $[t_n, t_{n+\frac{1}{2}}]$ and $[t_n, t_{n+1}]$. We split the integral approximation into four terms:

$$\int_{t_n}^{t_{n+1}} f(u(t),t)dt \approx \frac{\Delta t}{6}\left(f^n + 2\hat{f}^{n+\frac{1}{2}} + 2\tilde{f}^{n+\frac{1}{2}} + \bar{f}^{n+1}\right),$$

where $\hat{f}^{n+\frac{1}{2}}$, $\tilde{f}^{n+\frac{1}{2}}$, and $\bar{f}^{n+1}$ are approximations to $f^{n+\frac{1}{2}}$ and f^{n+1}, respectively, that can be based on already computed quantities. For $\hat{f}^{n+\frac{1}{2}}$ we can apply an approximation to $u^{n+\frac{1}{2}}$ using the Forward Euler method with step $\frac{1}{2}\Delta t$:

$$\hat{f}^{n+\frac{1}{2}} = f(u^n + \frac{1}{2}\Delta t f^n, t_{n+\frac{1}{2}}) \tag{3.48}$$

Since this gives us a prediction of $f^{n+\frac{1}{2}}$, we can for $\tilde{f}^{n+\frac{1}{2}}$ try a Backward Euler method to approximate $u^{n+\frac{1}{2}}$:

$$\tilde{f}^{n+\frac{1}{2}} = f(u^n + \frac{1}{2}\Delta t \hat{f}^{n+\frac{1}{2}}, t_{n+\frac{1}{2}})\,. \tag{3.49}$$

With $\tilde{f}^{n+\frac{1}{2}}$ as a hopefully good approximation to $f^{n+\frac{1}{2}}$, we can for the final term $\bar{f}^{n+1}$ use a Crank–Nicolson method on $[t_n, t_{n+1}]$ to approximate u^{n+1}:

$$\bar{f}^{n+1} = f(u^n + \Delta t \hat{f}^{n+\frac{1}{2}}, t_{n+1})\,. \tag{3.50}$$

[4] http://en.wikipedia.org/wiki/Simpson's_rule

We have now used the Forward and Backward Euler methods as well as the Crank–Nicolson method in the context of Simpson's rule. The hope is that the combination of these methods yields an overall time-stepping scheme from t_n to t_n+1 that is much more accurate than the $\mathcal{O}(\Delta t)$ and $\mathcal{O}(\Delta t^2)$ of the individual steps. This is indeed true: the overall accuracy is $\mathcal{O}(\Delta t^4)$!

To summarize, the 4th-order Runge–Kutta method becomes

$$u^{n+1} = u^n + \frac{\Delta t}{6}\left(f^n + 2\hat{f}^{n+\frac{1}{2}} + 2\tilde{f}^{n+\frac{1}{2}} + \bar{f}^{n+1}\right), \tag{3.51}$$

where the quantities on the right-hand side are computed from (3.48)–(3.50). Note that the scheme is fully explicit so there is never any need to solve linear or nonlinear algebraic equations. However, the stability is conditional and depends on f. There is a whole range of *implicit* Runge–Kutta methods that are unconditionally stable, but require solution of algebraic equations involving f at each time step.

The simplest way to explore more sophisticated methods for ODEs is to apply one of the many high-quality software packages that exist, as the next section explains.

3.2.9 The Odespy Software

A wide range of methods and software exist for solving (3.36). Many of the methods are accessible through a unified Python interface offered by the Odespy[5] [10] package. Odespy features simple Python implementations of the most fundamental schemes as well as Python interfaces to several famous packages for solving ODEs: ODEPACK[6], Vode[7], rkc.f[8], rkf45.f[9], as well as the ODE solvers in SciPy[10], SymPy[11], and odelab[12].

The code below illustrates the usage of Odespy the solving $u' = -au$, $u(0) = I$, $t \in (0, T]$, by the famous 4th-order Runge–Kutta method, using $\Delta t = 1$ and $N_t = 6$ steps:

```
def f(u, t):
    return -a*u

import odespy
import numpy as np

I = 1; a = 0.5; Nt = 6; dt = 1
solver = odespy.RK4(f)
solver.set_initial_condition(I)
t_mesh = np.linspace(0, Nt*dt, Nt+1)
u, t = solver.solve(t_mesh)
```

[5] https://github.com/hplgit/odespy
[6] https://computation.llnl.gov/casc/odepack/odepack_home.html
[7] https://computation.llnl.gov/casc/odepack/odepack_home.html
[8] http://www.netlib.org/ode/rkc.f
[9] http://www.netlib.org/ode/rkf45.f
[10] http://docs.scipy.org/doc/scipy/reference/generated/scipy.integrate.ode.html
[11] http://docs.sympy.org/dev/modules/mpmath/calculus/odes.html
[12] http://olivierverdier.github.com/odelab/

The previously listed methods for ODEs are all accessible in Odespy:

- the θ-rule: `ThetaRule`
- special cases of the θ-rule: `ForwardEuler`, `BackwardEuler`, `CrankNicolson`
- the 2nd- and 4th-order Runge–Kutta methods: `RK2` and `RK4`
- The BDF methods and the Adam-Bashforth methods: `Vode`, `Lsode`, `Lsoda`, `lsoda_scipy`
- The Leapfrog schemes: `Leapfrog` and `LeapfrogFiltered`

3.2.10 Example: Runge–Kutta Methods

Since all solvers have the same interface in Odespy, except for a potentially different set of parameters in the solvers' constructors, one can easily make a list of solver objects and run a loop for comparing a lot of solvers. The code below, found in complete form in `decay_odespy.py`[13], compares the famous Runge–Kutta methods of orders 2, 3, and 4 with the exact solution of the decay equation $u' = -au$. Since we have quite long time steps, we have included the only relevant θ-rule for large time steps, the Backward Euler scheme ($\theta = 1$), as well. Figure 3.1 shows the results.

```
import numpy as np
import matplotlib.pyplot as plt
import sys

def f(u, t):
    return -a*u

I = 1; a = 2; T = 6
dt = float(sys.argv[1]) if len(sys.argv) >= 2 else 0.75
Nt = int(round(T/dt))
t = np.linspace(0, Nt*dt, Nt+1)

solvers = [odespy.RK2(f),
           odespy.RK3(f),
           odespy.RK4(f),]

# BackwardEuler must use Newton solver to converge
# (Picard is default and leads to divergence)
solvers.append(
    odespy.BackwardEuler(f, nonlinear_solver='Newton'))
# Or tell BackwardEuler that it is a linear problem
solvers[-1] = odespy.BackwardEuler(f, f_is_linear=True,
                                   jac=lambda u, t: -a)]
legends = []
for solver in solvers:
    solver.set_initial_condition(I)
    u, t = solver.solve(t)

    plt.plot(t, u)
    plt.hold('on')
    legends.append(solver.__class__.__name__)
```

[13] http://tinyurl.com/ofkw6kc/genz/decay_odespy.py

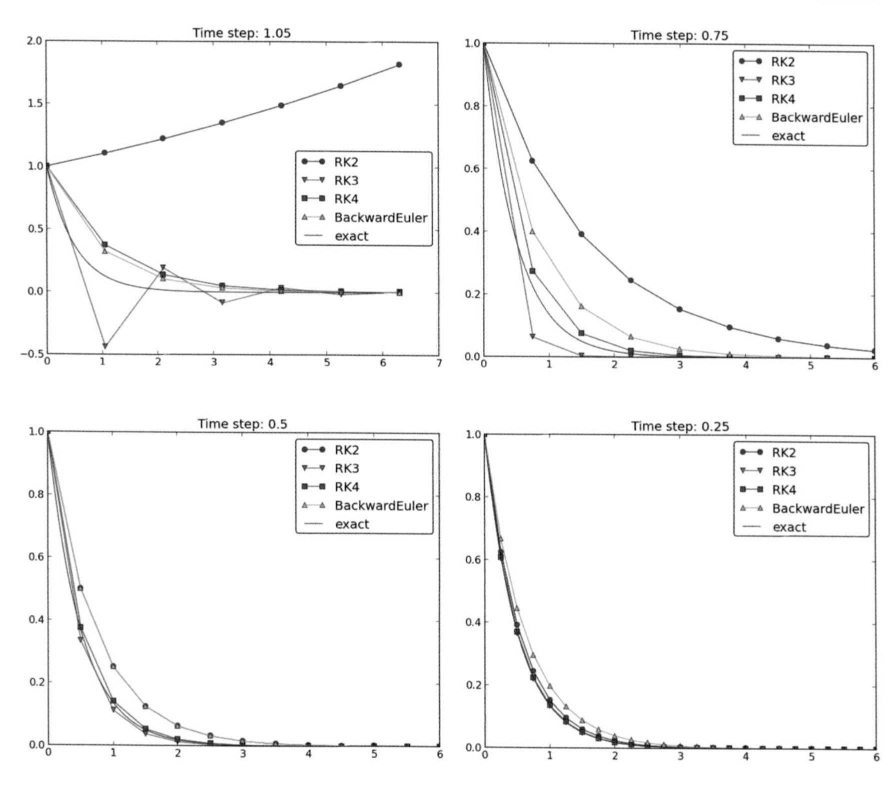

Fig. 3.1 Behavior of different schemes for the decay equation

```
# Compare with exact solution plotted on a very fine mesh
t_fine = np.linspace(0, T, 10001)
u_e = I*np.exp(-a*t_fine)
plt.plot(t_fine, u_e, '-') # avoid markers by specifying line type
legends.append('exact')

plt.legend(legends)
plt.title('Time step: %g' % dt)
plt.show()
```

With the `odespy.BackwardEuler` method we must either tell that the problem is linear and provide the Jacobian of $f(u,t)$, i.e., $\partial f/\partial u$, as the `jac` argument, or we have to assume that f is nonlinear, but then specify Newton's method as solver for the nonlinear equations (since the equations are linear, Newton's method will converge in one iteration). By default, `odespy.BackwardEuler` assumes a nonlinear problem to be solved by Picard iteration, but that leads to divergence in the present problem.

Visualization tip

We use Matplotlib for plotting here, but one could alternatively import `scitools.std` as `plt` instead. Plain use of Matplotlib as done here results in curves with different colors, which may be hard to distinguish on black-and-white paper. Using `scitools.std`, curves are automatically given colors *and* markers, thus making curves easy to distinguish on screen with colors and on black-and-white paper. The automatic adding of markers is normally a bad idea for a very fine mesh since all the markers get cluttered, but `scitools.std` limits the number of markers in such cases. For the exact solution we use a very fine mesh, but in the code above we specify the line type as a solid line (`-`), which means no markers and just a color to be automatically determined by the backend used for plotting (Matplotlib by default, but `scitools.std` gives the opportunity to use other backends to produce the plot, e.g., Gnuplot or Grace).

Also note the that the legends are based on the class names of the solvers, and in Python the name of the class type (as a string) of an object `obj` is obtained by `obj.__class__.__name__`.

The runs in Fig. 3.1 and other experiments reveal that the 2nd-order Runge–Kutta method (`RK2`) is unstable for $\Delta t > 1$ and decays slower than the Backward Euler scheme for large and moderate Δt (see Exercise 3.5 for an analysis). However, for fine $\Delta t = 0.25$ the 2nd-order Runge–Kutta method approaches the exact solution faster than the Backward Euler scheme. That is, the latter scheme does a better job for larger Δt, while the higher order scheme is superior for smaller Δt. This is a typical trend also for most schemes for ordinary and partial differential equations.

The 3rd-order Runge–Kutta method (`RK3`) also has artifacts in the form of oscillatory behavior for the larger Δt values, much like that of the Crank–Nicolson scheme. For finer Δt, the 3rd-order Runge–Kutta method converges quickly to the exact solution.

The 4th-order Runge–Kutta method (`RK4`) is slightly inferior to the Backward Euler scheme on the coarsest mesh, but is then clearly superior to all the other schemes. It is definitely the method of choice for all the tested schemes.

Remark about using the θ-rule in Odespy The Odespy package assumes that the ODE is written as $u' = f(u,t)$ with an f that is possibly nonlinear in u. The θ-rule for $u' = f(u,t)$ leads to

$$u^{n+1} = u^n + \Delta t \left(\theta f(u^{n+1}, t_{n+1}) + (1-\theta) f(u^n, t_n)\right),$$

which is a *nonlinear equation* in u^{n+1}. Odespy's implementation of the θ-rule (`ThetaRule`) and the specialized Backward Euler (`BackwardEuler`) and Crank–Nicolson (`CrankNicolson`) schemes must invoke iterative methods for solving the nonlinear equation in u^{n+1}. This is done even when f is linear in u, as in the model problem $u' = -au$, where we can easily solve for u^{n+1} by hand. Therefore, we need to specify use of Newton's method to solve the equations. (Odespy allows other methods than Newton's to be used, for instance Picard iteration, but that method is not suitable. The reason is that it applies the Forward Euler scheme to generate a start value for the iterations. Forward Euler may give very wrong solutions for large Δt values. Newton's method, on the other hand, is insensitive to the start value in *linear problems*.)

3.2.11 Example: Adaptive Runge–Kutta Methods

Odespy also offers solution methods that can adapt the size of Δt with time to match a desired accuracy in the solution. Intuitively, small time steps will be chosen in areas where the solution is changing rapidly, while larger time steps can be used where the solution is slowly varying. Some kind of *error estimator* is used to adjust the next time step at each time level.

A very popular adaptive method for solving ODEs is the Dormand-Prince Runge–Kutta method of order 4 and 5. The 5th-order method is used as a reference solution and the difference between the 4th- and 5th-order methods is used as an indicator of the error in the numerical solution. The Dormand-Prince method is the default choice in MATLAB's widely used `ode45` routine.

We can easily set up Odespy to use the Dormand-Prince method and see how it selects the optimal time steps. To this end, we request only one time step from $t = 0$ to $t = T$ and ask the method to compute the necessary non-uniform time mesh to meet a certain error tolerance. The code goes like

```
import odespy
import numpy as np
import decay_mod
import sys
#import matplotlib.pyplot as plt
import scitools.std as plt

def f(u, t):
    return -a*u

def u_exact(t):
    return I*np.exp(-a*t)

I = 1; a = 2; T = 5
tol = float(sys.argv[1])
solver = odespy.DormandPrince(f, atol=tol, rtol=0.1*tol)

Nt = 1  # just one step - let the scheme find
        # its intermediate points
t_mesh = np.linspace(0, T, Nt+1)
t_fine = np.linspace(0, T, 10001)

solver.set_initial_condition(I)
u, t = solver.solve(t_mesh)

# u and t will only consist of [I, u^Nt] and [0,T]
# solver.u_all and solver.t_all contains all computed points
plt.plot(solver.t_all, solver.u_all, 'ko')
plt.hold('on')
plt.plot(t_fine, u_exact(t_fine), 'b-')
plt.legend(['tol=%.0E' % tol, 'exact'])
plt.savefig('tmp_odespy_adaptive.png')
plt.show()
```

Running four cases with tolerances 10^{-1}, 10^{-3}, 10^{-5}, and 10^{-7}, gives the results in Fig. 3.2. Intuitively, one would expect denser points in the beginning of the decay and larger time steps when the solution flattens out.

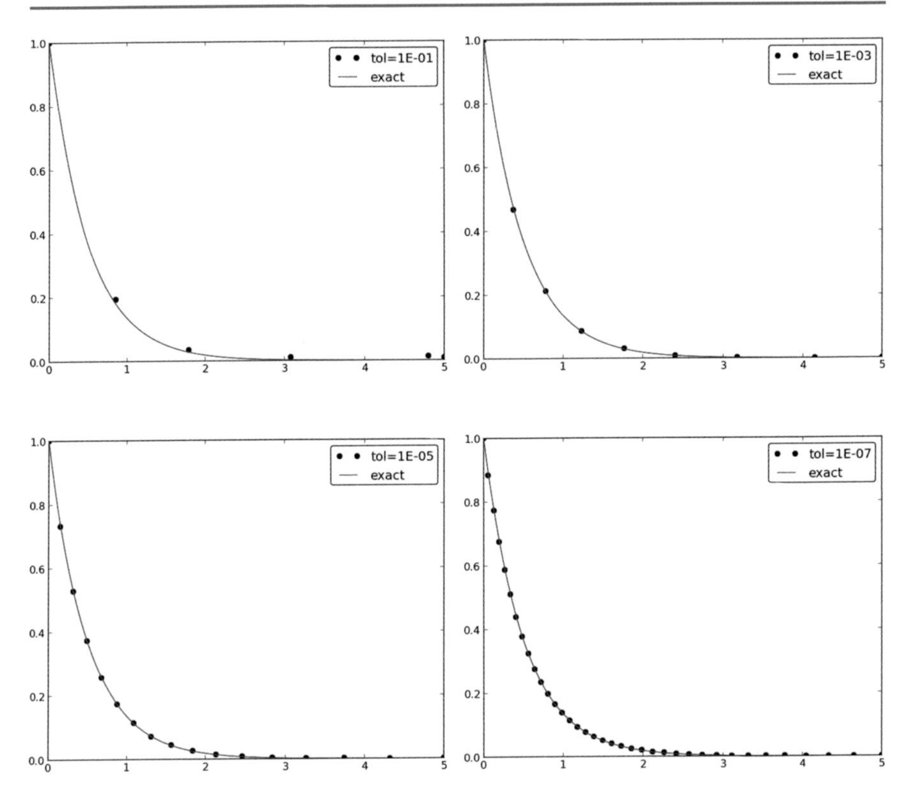

Fig. 3.2 Choice of adaptive time mesh by the Dormand-Prince method for different tolerances

3.3 Exercises

Exercise 3.1: Experiment with precision in tests and the size of *u*
It is claimed in Sect. 3.1.5 that most numerical methods will reproduce a linear exact solution to machine precision. Test this assertion using the test function `test_linear_solution` in the `decay_vc.py`[14] program. Vary the parameter `c` from very small, via `c=1` to many larger values, and print out the maximum difference between the numerical solution and the exact solution. What is the relevant value of the tolerance in the float comparison in each case?

Filename: `test_precision`.

Exercise 3.2: Implement the 2-step backward scheme
Implement the 2-step backward method (3.39) for the model $u'(t) = -a(t)u(t) + b(t)$, $u(0) = I$. Allow the first step to be computed by either the Backward Euler scheme or the Crank–Nicolson scheme. Verify the implementation by choosing $a(t)$ and $b(t)$ such that the exact solution is linear in t (see Sect. 3.1.5). Show mathematically that a linear solution is indeed a solution of the discrete equations.

Compute convergence rates (see Sect. 3.1.6) in a test case using $a = \text{const}$ and $b = 0$, where we easily have an exact solution, and determine if the choice

[14] http://tinyurl.com/ofkw6kc/genz/decay_vc.py

of a first-order scheme (Backward Euler) for the first step has any impact on the overall accuracy of this scheme. The expected error goes like $\mathcal{O}(\Delta t^2)$.

Filename: `decay_backward2step`.

Exercise 3.3: Implement the 2nd-order Adams–Bashforth scheme

Implement the 2nd-order Adams–Bashforth method (3.46) for the decay problem $u' = -a(t)u + b(t)$, $u(0) = I$, $t \in (0, T]$. Use the Forward Euler method for the first step such that the overall scheme is explicit. Verify the implementation using an exact solution that is linear in time. Analyze the scheme by searching for solutions $u^n = A^n$ when $a = \text{const}$ and $b = 0$. Compare this second-order scheme to the Crank–Nicolson scheme.

Filename: `decay_AdamsBashforth2`.

Exercise 3.4: Implement the 3rd-order Adams–Bashforth scheme

Implement the 3rd-order Adams–Bashforth method (3.47) for the decay problem $u' = -a(t)u + b(t)$, $u(0) = I$, $t \in (0, T]$. Since the scheme is explicit, allow it to be started by two steps with the Forward Euler method. Investigate experimentally the case where $b = 0$ and a is a constant: Can we have oscillatory solutions for large Δt?

Filename: `decay_AdamsBashforth3`.

Exercise 3.5: Analyze explicit 2nd-order methods

Show that the schemes (3.44) and (3.45) are identical in the case $f(u, t) = -a$, where $a > 0$ is a constant. Assume that the numerical solution reads $u^n = A^n$ for some unknown amplification factor A to be determined. Find A and derive stability criteria. Can the scheme produce oscillatory solutions of $u' = -au$? Plot the numerical and exact amplification factor.

Filename: `decay_RK2_Taylor2`.

Project 3.6: Implement and investigate the Leapfrog scheme

A Leapfrog scheme for the ODE $u'(t) = -a(t)u(t) + b(t)$ is defined by

$$[D_{2t}u = -au + b]^n \,. \tag{3.52}$$

A separate method is needed to compute u^1. The Forward Euler scheme is a possible candidate.

a) Implement the Leapfrog scheme for the model equation. Plot the solution in the case $a = 1$, $b = 0$, $I = 1$, $\Delta t = 0.01$, $t \in [0, 4]$. Compare with the exact solution $u_e(t) = e^{-t}$.
b) Show mathematically that a linear solution in t fulfills the Forward Euler scheme for the first step and the Leapfrog scheme for the subsequent steps. Use this linear solution to verify the implementation, and automate the verification through a test function.

Hint It can be wise to automate the calculations such that it is easy to redo the calculations for other types of solutions. Here is a possible `sympy` function that takes a symbolic expression u (implemented as a Python function of t), fits the b term, and checks if u fulfills the discrete equations:

```
import sympy as sym

def analyze(u):
    t, dt, a = sym.symbols('t dt a')

    print 'Analyzing u_e(t)=%s' % u(t)
    print 'u(0)=%s' % u(t).subs(t, 0)

    # Fit source term to the given u(t)
    b = sym.diff(u(t), t) + a*u(t)
    b = sym.simplify(b)
    print 'Source term b:', b

    # Residual in discrete equations; Forward Euler step
    R_step1 = (u(t+dt) - u(t))/dt + a*u(t) - b
    R_step1 = sym.simplify(R_step1)
    print 'Residual Forward Euler step:', R_step1

    # Residual in discrete equations; Leapfrog steps
    R = (u(t+dt) - u(t-dt))/(2*dt) + a*u(t) - b
    R = sym.simplify(R)
    print 'Residual Leapfrog steps:', R

def u_e(t):
    return c*t + I

analyze(u_e)
# or short form: analyze(lambda t: c*t + I)
```

c) Show that a second-order polynomial in t cannot be a solution of the discrete equations. However, if a Crank–Nicolson scheme is used for the first step, a second-order polynomial solves the equations exactly.
d) Create a manufactured solution $u(t) = \sin(t)$ for the ODE $u' = -au + b$. Compute the convergence rate of the Leapfrog scheme using this manufactured solution. The expected convergence rate of the Leapfrog scheme is $\mathcal{O}(\Delta t^2)$. Does the use of a 1st-order method for the first step impact the convergence rate?
e) Set up a set of experiments to demonstrate that the Leapfrog scheme (3.52) is associated with numerical artifacts (instabilities). Document the main results from this investigation.
f) Analyze and explain the instabilities of the Leapfrog scheme (3.52):
    1. Choose $a = \text{const}$ and $b = 0$. Assume that an exact solution of the discrete equations has the form $u^n = A^n$, where A is an amplification factor to be determined. Derive an equation for A by inserting $u^n = A^n$ in the Leapfrog scheme.
    2. Compute A either by hand and/or with the aid of sympy. The polynomial for A has two roots, A_1 and A_2. Let u^n be a linear combination $u^n = C_1A_1^n + C_2A_2^n$.
    3. Show that one of the roots is the reason for instability.
    4. Compare A with the exact expression, using a Taylor series approximation.
    5. How can C_1 and C_2 be determined?
g) Since the original Leapfrog scheme is unconditionally unstable as time grows, it demands some stabilization. This can be done by filtering, where we first find

u^{n+1} from the original Leapfrog scheme and then replace u^n by $u^n + \gamma(u^{n-1} - 2u^n + u^{n+1})$, where γ can be taken as 0.6. Implement the filtered Leapfrog scheme and check that it can handle tests where the original Leapfrog scheme is unstable.

Filename: `decay_leapfrog`.

Problem 3.7: Make a unified implementation of many schemes
Consider the linear ODE problem $u'(t) = -a(t)u(t) + b(t)$, $u(0) = I$. Explicit schemes for this problem can be written in the general form

$$u^{n+1} = \sum_{j=0}^{m} c_j u^{n-j}, \tag{3.53}$$

for some choice of $c_0, \ldots, c_m$. Find expressions for the c_j coefficients in case of the θ-rule, the three-level backward scheme, the Leapfrog scheme, the 2nd-order Runge–Kutta method, and the 3rd-order Adams–Bashforth scheme.

Make a class `ExpDecay` that implements the general updating formula (3.53). The formula cannot be applied for $n < m$, and for those n values, other schemes must be used. Assume for simplicity that we just repeat Crank–Nicolson steps until (3.53) can be used. Use a subclass to specify the list $c_0, \ldots, c_m$ for a particular method, and implement subclasses for all the mentioned schemes. Verify the implementation by testing with a linear solution, which should be exactly reproduced by all methods.

Filename: `decay_schemes_unified`.

Models

4

This chapter presents many mathematical models that all end up with ODEs of the type $u' = -au + b$. The applications are taken from biology, finance, and physics, and cover population growth or decay, interacting predator-prey populations, compound interest and inflation, radioactive decay, chemical and biochemical reaction, spreading of diseases, cooling of objects, compaction of geological media, pressure variations in the atmosphere, viscoelastic response in materials, and air resistance on falling or rising bodies.

Before we turn to the applications, however, we take a brief look at the technique of scaling, which is so useful in many applications.

4.1 Scaling

Real applications of a model $u' = -au + b$ will often involve a lot of parameters in the expressions for a and b. It can be quite a challenge to find relevant values of all parameters. In simple problems, however, it turns out that it is not always necessary to estimate all parameters because we can lump them into one or a few *dimensionless* numbers by using a very attractive technique called scaling. It simply means to stretch the u and t axis in the present problem – and suddenly all parameters in the problem are lumped into one parameter if $b \neq 0$ and no parameter when $b = 0$!

4.1.1 Dimensionless Variables

Scaling means that we introduce a new function $\bar{u}(\bar{t})$, with

$$\bar{u} = \frac{u - u_m}{u_c}, \quad \bar{t} = \frac{t}{t_c},$$

where u_m is a characteristic value of u, u_c is a characteristic size of the range of u values, and t_c is a characteristic size of the range of t where u shows significant variation. Choosing u_m, u_c, and t_c is not always easy and is often an art in complicated problems. We just state one choice first:

$$u_c = I, \quad u_m = b/a, \quad t_c = 1/a\,.$$

H.P. Langtangen, *Finite Difference Computing with Exponential Decay Models*,
Lecture Notes in Computational Science and Engineering 110,
DOI 10.1007/978-3-319-29439-1_4

Inserting $u = u_m + u_c\bar{u}$ and $t = t_c\bar{t}$ in the problem $u' = -au + b$, assuming a and b are constants, results (after some algebra) in the *scaled problem*

$$\frac{d\bar{u}}{d\bar{t}} = -\bar{u}, \quad \bar{u}(0) = 1 - \beta,$$

where

$$\beta = \frac{b}{Ia}.$$

4.1.2 Dimensionless Numbers

The parameter β is a dimensionless number. From the equation we see that b must have the same unit as the term au. The initial condition I must have the same unit as u, so Ia has the same unit as b, making the fraction $b/(Ia)$ dimensionless.

An important observation is that $\bar{u}$ depends on $\bar{t}$ and β. That is, only the special combination of $b/(Ia)$ matters, not what the individual values of b, a, and I are. The original unscaled function u depends on t, b, a, and I.

A second observation is striking: if $b = 0$, the scaled problem is independent of a and I! In practice this means that we can perform a single numerical simulation of the scaled problem and recover the solution of any problem for a given a and I by stretching the axis in the plot: $u = I\bar{u}$ and $t = \bar{t}/a$. For $b \neq 0$, we simulate the scaled problem for a few β values and recover the physical solution u by translating and stretching the u axis and stretching the t axis.

In general, scaling combines the parameters in a problem to a set of dimensionless parameters. The number of dimensionless parameters is usually much smaller than the number of original parameters. Section 4.11 presents an example where 11 parameters are reduced to one!

4.1.3 A Scaling for Vanishing Initial Condition

The scaling breaks down if $I = 0$. In that case we may choose $u_m = 0$, $u_c = b/a$, and $t_c = 1/b$, resulting in a slightly different scaled problem:

$$\frac{d\bar{u}}{d\bar{t}} = 1 - \bar{u}, \quad \bar{u}(0) = 0.$$

As with $b = 0$, the case $I = 0$ has a scaled problem with no physical parameters!

It is common to drop the bars after scaling and write the scaled problem as $u' = -u$, $u(0) = 1 - \beta$, or $u' = 1 - u$, $u(0) = 0$. Any implementation of the problem $u' = -au + b$, $u(0) = I$, can be reused for the scaled problem by setting $a = 1$, $b = 0$, and $I = 1 - \beta$ in the code, if $I \neq 0$, or one sets $a = 1$, $b = 1$, and $I = 0$ when the physical I is zero. Falling bodies in fluids, as described in Sect. 4.11, involves $u' = -au + b$ with seven physical parameters. All these vanish in the scaled version of the problem if we start the motion from rest!

Many more details about scaling are presented in the author's book *Scaling of Differential Equations* [9].

4.2 Evolution of a Population

4.2.1 Exponential Growth

Let N be the number of individuals in a population occupying some spatial domain. Despite N being an integer in this problem, we shall compute with N as a real number and view $N(t)$ as a continuous function of time. The basic model assumption is that in a time interval Δt the number of newcomers to the populations (newborns) is proportional to N, with proportionality constant $\bar{b}$. The amount of newcomers will increase the population and result in

$$N(t + \Delta t) = N(t) + \bar{b}N(t) .$$

It is obvious that a long time interval Δt will result in more newcomers and hence a larger $\bar{b}$. Therefore, we introduce $b = \bar{b}/\Delta t$: the number of newcomers per unit time and per individual. We must then multiply b by the length of the time interval considered and by the population size to get the total number of new individuals, $b\Delta tN$.

If the number of removals from the population (deaths) is also proportional to N, with proportionality constant $d\Delta t$, the population evolves according to

$$N(t + \Delta t) = N(t) + b\Delta tN(t) - d\Delta tN(t) .$$

Dividing by Δt and letting $\Delta t \to 0$, we get the ODE

$$N' = (b - d)N, \quad N(0) = N_0 . \tag{4.1}$$

In a population where the death rate (d) is larger than then newborn rate (b), $b-d < 0$, and the population experiences exponential decay rather than exponential growth.

In some populations there is an immigration of individuals into the spatial domain. With I individuals coming in per time unit, the equation for the population change becomes

$$N(t + \Delta t) = N(t) + b\Delta tN(t) - d\Delta tN(t) + \Delta tI .$$

The corresponding ODE reads

$$N' = (b - d)N + I, \quad N(0) = N_0 . \tag{4.2}$$

Emigration is also modeled by this I term if we just change its sign: $I < 0$. So, the I term models migration in and out of the domain in general.

Some simplification arises if we introduce a fractional measure of the population: $u = N/N_0$ and set $r = b - d$. The ODE problem now becomes

$$u' = ru + f, \quad u(0) = 1, \tag{4.3}$$

where $f = I/N_0$ measures the net immigration per time unit as the fraction of the initial population. Very often, r is approximately constant, but f is usually a function of time.

4.2.2 Logistic Growth

The growth rate r of a population decreases if the environment has limited resources. Suppose the environment can sustain at most $N_{\max}$ individuals. We may then assume that the growth rate approaches zero as N approaches $N_{\max}$, i.e., as u approaches $M = N_{\max}/N_0$. The simplest possible evolution of r is then a linear function: $r(t) = \varrho(1-u(t)/M)$, where ϱ is the initial growth rate when the population is small relative to the maximum size and there is enough resources. Using this $r(t)$ in (4.3) results in the *logistic model* for the evolution of a population (assuming for the moment that $f = 0$):

$$u' = \varrho(1 - u/M)u, \quad u(0) = 1\,. \tag{4.4}$$

Initially, u will grow at rate ϱ, but the growth will decay as u approaches M, and then there is no more change in u, causing $u \to M$ as $t \to \infty$. Note that the logistic equation $u' = \varrho(1 - u/M)u$ is *nonlinear* because of the quadratic term $-u^2\varrho/M$.

4.3 Compound Interest and Inflation

Say the annual interest rate is r percent and that the bank adds the interest once a year to your investment. If u^n is the investment in year n, the investment in year u^{n+1} grows to

$$u^{n+1} = u^n + \frac{r}{100}u^n\,.$$

In reality, the interest rate is added every day. We therefore introduce a parameter m for the number of periods per year when the interest is added. If n counts the periods, we have the fundamental model for compound interest:

$$u^{n+1} = u^n + \frac{r}{100m}u^n\,. \tag{4.5}$$

This model is a *difference equation*, but it can be transformed to a continuous differential equation through a limit process. The first step is to derive a formula for the growth of the investment over a time t. Starting with an investment u^0, and assuming that r is constant in time, we get

$$\begin{aligned} u^{n+1} &= \left(1 + \frac{r}{100m}\right)u^n \\ &= \left(1 + \frac{r}{100m}\right)^2 u^{n-1} \\ &\ \ \vdots \\ &= \left(1 + \frac{r}{100m}\right)^{n+1} u^0 \end{aligned}$$

Introducing time t, which here is a real-numbered counter for years, we have that $n = mt$, so we can write

$$u^{mt} = \left(1 + \frac{r}{100m}\right)^{mt} u^0\,.$$

The second step is to assume *continuous compounding*, meaning that the interest is added continuously. This implies $m \to \infty$, and in the limit one gets the formula

$$u(t) = u_0 e^{rt/100}, \tag{4.6}$$

which is nothing but the solution of the ODE problem

$$u' = \frac{r}{100}u, \quad u(0) = u_0 \,. \tag{4.7}$$

This is then taken as the ODE model for compound interest if $r > 0$. However, the reasoning applies equally well to inflation, which is just the case $r < 0$. One may also take the r in (4.7) as the net growth of an investment, where r takes both compound interest and inflation into account. Note that for real applications we must use a time-dependent r in (4.7).

Introducing $a = \frac{r}{100}$, continuous inflation of an initial fortune I is then a process exhibiting exponential decay according to

$$u' = -au, \quad u(0) = I \,.$$

4.4 Newton's Law of Cooling

When a body at some temperature is placed in a cooling environment, experience shows that the temperature falls rapidly in the beginning, and then the change in temperature levels off until the body's temperature equals that of the surroundings. Newton carried out some experiments on cooling hot iron and found that the temperature evolved as a "geometric progression at times in arithmetic progression", meaning that the temperature decayed exponentially. Later, this result was formulated as a differential equation: the rate of change of the temperature in a body is proportional to the temperature difference between the body and its surroundings. This statement is known as *Newton's law of cooling*, which mathematically can be expressed as

$$\frac{dT}{dt} = -k(T - T_s), \tag{4.8}$$

where T is the temperature of the body, T_s is the temperature of the surroundings (which may be time-dependent), t is time, and k is a positive constant. Equation (4.8) is primarily viewed as an empirical law, valid when heat is efficiently convected away from the surface of the body by a flowing fluid such as air at constant temperature T_s. The *heat transfer coefficient* k reflects the transfer of heat from the body to the surroundings and must be determined from physical experiments.

The cooling law (4.8) needs an initial condition $T(0) = T_0$.

4.5 Radioactive Decay

An atomic nucleus of an unstable atom may lose energy by emitting ionizing particles and thereby be transformed to a nucleus with a different number of protons and neutrons. This process is known as radioactive decay[1]. Actually, the process

[1] http://en.wikipedia.org/wiki/Radioactive_decay

is stochastic when viewed for a single atom, because it is impossible to predict exactly when a particular atom emits a particle. Nevertheless, with a large number of atoms, N, one may view the process as deterministic and compute the mean behavior of the decay. Below we reason intuitively about an ODE for the mean behavior. Thereafter, we show mathematically that a detailed stochastic model for single atoms leads to the same mean behavior.

4.5.1 Deterministic Model

Suppose at time t, the number of the original atom type is $N(t)$. A basic model assumption is that the transformation of the atoms of the original type in a small time interval Δt is proportional to N, so that

$$N(t + \Delta t) = N(t) - a\Delta t N(t),$$

where $a > 0$ is a constant. The proportionality factor is $a\Delta t$, i.e., proportional to Δt since a longer time interval will produce more transformations. We can introduce $u = N(t)/N(0)$, divide by Δt, and let $\Delta t \to 0$:

$$\lim_{r\to 0} N_0 \frac{u(t + \Delta t) - u(t)}{\Delta t} = -aN_0 u(t)\,.$$

The left-hand side is the derivative of u. Dividing by the N_0 gives the following ODE for u:

$$u' = -au, \quad u(0) = 1\,. \tag{4.9}$$

The parameter a can for a given nucleus be expressed through the *half-life* $t_{1/2}$, which is the time taken for the decay to reduce the initial amount by one half, i.e., $u(t_{1/2}) = 0.5$. With $u(t) = e^{-at}$, we get $t_{1/2} = a^{-1}\ln 2$ or $a = \ln 2/t_{1/2}$.

4.5.2 Stochastic Model

Originally, we have N_0 atoms. Up to some particular time t, each atom may either have decayed or not. If not, they have "survived". We want to count how many original atoms that have survived. The survival of a single atom at time t is a random event. Since there are only two outcomes, survival or decay, we have a Bernoulli trial[2]. Let p be the probability of survival (implying that the probability of decay is $1 - p$). If each atom survives independently of the others, and the probability of survival is the same for every atom, we have N_0 Bernoulli trials, known as a *binomial experiment* from probability theory. The probability $P(N)$ that N out of the N_0 atoms have survived at time t is then given by the famous *binomial distribution*

$$P(N) = \frac{N_0!}{N!(N_0 - N)!} p^N (1 - p)^{N_0 - N}\,.$$

The mean (or expected) value $\mathrm{E}[P]$ of $P(N)$ is known to be $N_0 p$.

[2] http://en.wikipedia.org/wiki/Bernoulli_trial

It remains to estimate p. Let the interval $[0, t]$ be divided into m small subintervals of length Δt. We make the assumption that the probability of decay of a single atom in an interval of length Δt is $\tilde{p}$, and that this probability is proportional to Δt: $\tilde{p} = \lambda \Delta t$ (it sounds natural that the probability of decay increases with Δt). The corresponding probability of survival is $1 - \lambda \Delta t$. Believing that λ is independent of time, we have, for each interval of length Δt, a Bernoulli trial: the atom either survives or decays in that interval. Now, p should be the probability that the atom survives in all the intervals, i.e., that we have m successful Bernoulli trials in a row and therefore

$$p = (1 - \lambda \Delta t)^m .$$

The expected number of atoms of the original type at time t is

$$\mathrm{E}[P] = N_0 p = N_0(1 - \lambda \Delta t)^m, \quad m = t/\Delta t . \tag{4.10}$$

To see the relation between the two types of Bernoulli trials and the ODE above, we go to the limit $\Delta t \to 0$, $m \to \infty$. It is possible to show that

$$p = \lim_{m\to\infty} (1 - \lambda \Delta t)^m = \lim_{m\to\infty} \left(1 - \lambda \frac{t}{m}\right)^m = e^{-\lambda t}$$

This is the famous exponential waiting time (or arrival time) distribution for a Poisson process in probability theory (obtained here, as often done, as the limit of a binomial experiment). The probability of decay, or more precisely that at least one atom has undergone a transition, is $1 - p = 1 - e^{-\lambda t}$. This is the exponential distribution[3]. The limit means that m is very large, hence Δt is very small, and $\tilde{p} = \lambda \Delta t$ is very small since the intensity of the events, λ, is assumed finite. This situation corresponds to a very small probability that an atom will decay in a very short time interval, which is a reasonable model. The same model occurs in lots of different applications, e.g., when waiting for a taxi, or when finding defects along a rope.

4.5.3 Relation Between Stochastic and Deterministic Models

With $p = e^{-\lambda t}$ we get the expected number of original atoms at t as $N_0 p = N_0 e^{-\lambda t}$, which is exactly the solution of the ODE model $N' = -\lambda N$. This also gives an interpretation of a via λ or vice versa. Our important finding here is that the ODE model captures the mean behavior of the underlying stochastic model. This is, however, not always the common relation between microscopic stochastic models and macroscopic "averaged" models.

Also of interest, is that a Forward Euler discretization of $N' = -\lambda N$, $N(0) = N_0$, gives $N^m = N_0(1 - \lambda \Delta t)^m$ at time $t_m = m\Delta t$, which is exactly the expected value of the stochastic experiment with N_0 atoms and m small intervals of length Δt, where each atom can decay with probability $\lambda \Delta t$ in an interval.

A fundamental question is how accurate the ODE model is. The underlying stochastic model fluctuates around its expected value. A measure of the fluctuations

[3] http://en.wikipedia.org/wiki/Exponential_distribution

is the standard deviation of the binomial experiment with N_0 atoms, which can be shown to be $\mathrm{Std}[P] = \sqrt{N_0 p(1-p)}$. Compared to the size of the expectation, we get the normalized standard deviation

$$\frac{\sqrt{\mathrm{Var}(P)}}{\mathrm{E}[P]} = N_0^{-1/2}\sqrt{p^{-1}-1} = N_0^{-1/2}\sqrt{(1-e^{-\lambda t})^{-1}-1} \approx (N_0\lambda t)^{-1/2},$$

showing that the normalized fluctuations are very small if N_0 is very large, which is usually the case.

4.5.4 Generalization of the Radioactive Decay Modeling

The modeling in Sect. 4.5 is in fact very general, despite a focus on a particular physical process. We may instead of atoms and decay speak about a set of *items*, where each item can undergo a stochastic *transition* from one state to another. In Sect. 4.6 the item is a molecule and the transition is a chemical reaction, while in Sect. 4.7 the item is an ill person and the transition is recovering from the illness (or an immune person who loses her immunity).

From the modeling in Sect. 4.5 we can establish a deterministic model for a large number of items and a stochastic model for an arbitrary number of items, even a single one. The stochastic model has a parameter λ reflecting the probability that a transition takes place in a time interval of unit length (or equivalently, that the probability is $\lambda \Delta t$ for a transition during a time interval of length Δt). The probability of making a transition before time t is given by

$$F(t) = 1 - e^{-\lambda t}.$$

The corresponding probability density is $f(t) = F'(t) = e^{-\lambda t}$. The expected value of $F(t)$, i.e., the expected time to transition, is λ^{-1}. This interpretation of λ makes it easy to measure its value: just carry out a large number of experiments, measure the time to transition, and take one over the average of these times as an estimate of λ. The variance is λ^{-2}.

The deterministic model counts how many items, $N(t)$, that have undergone the transition (on average), and $N(t)$ is governed by the ODE

$$N' = -\lambda N(t), \quad N(0) = N_0.$$

4.6 Chemical Kinetics

4.6.1 Irreversible Reaction of Two Substances

Consider two chemical substances, A and B, and a chemical reaction that turns A into B. In a small time interval, some of the molecules of type A are transformed into molecules of B. This process is, from a mathematical modeling point of view, equivalent to the radioactive decay process described in the previous section. We can therefore apply the same modeling approach. If N_A is the number of molecules of substance A, we have that N_A is governed by the differential equation

$$\frac{dN_A}{dt} = -kN_A,$$

where (the constant) k is called the *rate constant* of the reaction. Rather than using the number of molecules, we use the *concentration* of molecules: $[A](t) = N_A(t)/N_A(0)$. We see that $d[A]/dt = N_A(0)^{-1}dN_A/dt$. Replacing N_A by $[A]$ in the equation for N_A leads to the equation for the concentration $[A]$:

$$\frac{d[A]}{dt} = -k[A], \quad t \in (0,T],\ [A](0) = 1\,. \tag{4.11}$$

Since substance A is transformed to substance B, we have that the concentration of $[B]$ grows by the loss of $[A]$:

$$\frac{d[B]}{dt} = k[A], \quad [B](0) = 0\,.$$

The mathematical model can either be (4.11) or the system

$$\frac{d[A]}{dt} = -k[A], \qquad t \in (0,T] \tag{4.12}$$

$$\frac{d[B]}{dt} = k[A], \qquad t \in (0,T] \tag{4.13}$$

$$[A](0) = 1, \tag{4.14}$$

$$[B](0) = 0\,. \tag{4.15}$$

This reaction is known as a *first-order reaction*, where each molecule of A makes an independent decision about whether to complete the reaction, i.e., independent of what happens to any other molecule.

An n-th order reaction is modeled by

$$\frac{d[A]}{dt} = -k[A]^n, \tag{4.16}$$

$$\frac{d[B]}{dt} = k[A]^n, \tag{4.17}$$

for $t \in (0,T]$ with initial conditions $[A](0) = 1$ and $[B](0) = 0$. Here, n can be a real number, but is most often an integer. Note that the sum of the concentrations is constant since

$$\frac{d[A]}{dt} + \frac{d[B]}{dt} = 0 \quad \Rightarrow \quad [A](t) + [B](t) = \text{const} = [A](0) + [B](0) = 1 + 0\,.$$

4.6.2 Reversible Reaction of Two Substances

Let the chemical reaction turn substance A into B and substance B into A. The rate of change of $[A]$ has then two contributions: a loss $k_A[A]$ and a gain $k_B[B]$:

$$\frac{d[A]}{dt} = -k_A[A] + k_B[B], \quad t \in (0,T],\ [A](0) = A_0\,. \tag{4.18}$$

Similarly for substance B,

$$\frac{d[B]}{dt} = k_A[A] - k_B[B], \quad t \in (0, T],\ [B](0) = B_0\,. \tag{4.19}$$

This time we have allowed for arbitrary initial concentrations. Again,

$$\frac{d[A]}{dt} + \frac{d[B]}{dt} = 0 \quad \Rightarrow \quad [A](t) + [B](t) = A_0 + B_0\,.$$

4.6.3 Irreversible Reaction of Two Substances into a Third

Now we consider two chemical substances, A and B, reacting with each other and producing a substance C. In a small time interval Δt, molecules of type A and B are occasionally colliding, and in some of the collisions, a chemical reaction occurs, which turns A and B into a molecule of type C. (More generally, M_A molecules of A and M_B molecules of B react to form M_C molecules of C.) The number of possible pairings, and thereby collisions, of A and B is $N_A N_B$, where N_A is the number of molecules of A, and N_B is the number of molecules of B. A fraction k of these collisions, $\hat{k}\Delta t N_A N_B$, features a chemical reaction and produce N_C molecules of C. The fraction is thought to be proportional to Δt: considering a twice as long time interval, twice as many molecules collide, and twice as many reactions occur. The increase in molecules of substance C is now found from the reasoning

$$N_C(t + \Delta t) = N_C(t) + \hat{k}\Delta t N_A N_B\,.$$

Dividing by Δt,

$$\frac{N_C(t + \Delta t) - N_C(t)}{\Delta t} = \hat{k} N_A N_B,$$

and letting $\Delta t \to 0$, gives the differential equation

$$\frac{dN_C}{dt} = \hat{k} N_A N_B\,.$$

(This equation is known as the important law of mass action[4] discovered by the Norwegian scientists Cato M. Guldberg and Peter Waage. A more general form of the right-hand side is $\hat{k} N_A^{\alpha} N_B^{\beta}$. All the constants $\hat{k}$, α, and β must be determined from experiments.)

Working instead with concentrations, we introduce $[C](t) = N_C(t)/N_C(0)$, with similar definitions for $[A]$ and $[B]$ we get

$$\frac{d[C]}{dt} = k[A][B]\,. \tag{4.20}$$

[4] https://en.wikipedia.org/wiki/Law_of_mass_action

The constant k is related to $\hat{k}$ by $k = \hat{k}N_A(0)N_B(0)/N_C(0)$. The gain in C is a loss of A and B:

$$\frac{d[A]}{dt} = -k[A][B], \tag{4.21}$$

$$\frac{d[B]}{dt} = -k[A][B]. \tag{4.22}$$

4.6.4 A Biochemical Reaction

A common reaction (known as Michaelis–Menten kinetics[5]) turns a substrate S into a product P with aid of an enzyme E. The reaction is a two-stage process: first S and E reacts to form a complex ES, where the enzyme and substrate are bound to each other, and then ES is turned into E and P. In the first stage, S and E react to produce a growth of ES according to the law of mass action:

$$\frac{d[S]}{dt} = -k_+[E][S],$$

$$\frac{d[ES]}{dt} = k_+[E][S].$$

The complex ES reacts and produces the product P at rate $-k_v[ES]$ and E at rate $-k_-[ES]$. The total set of reactions can then be expressed by

$$\frac{d[ES]}{dt} = k_+[E][S] - k_v[ES] - k_-[ES], \tag{4.23}$$

$$\frac{d[P]}{dt} = k_v[ES], \tag{4.24}$$

$$\frac{d[S]}{dt} = -k_+[E][S] + k_-[ES], \tag{4.25}$$

$$\frac{d[E]}{dt} = -k_+[E][S] + k_-[ES] + k_v[ES]. \tag{4.26}$$

The initial conditions are $[ES](0) = [P](0) = 0$, and $[S] = S_0$, $[E] = E_0$. The constants k_+, k_-, and k_v must be determined from experiments.

4.7 Spreading of Diseases

The modeling of spreading of diseases is very similar to the modeling of chemical reactions in Sect. 4.6. The field of epidemiology speaks about susceptibles: people who can get a disease; infectives: people who are infected and can infect susceptibles; and recovered: people who have recovered from the disease and become immune. Three categories are accordingly defined: S for susceptibles, I for infectives, and R for recovered. The number in each category is tracked by the functions $S(t)$, $I(t)$, and $R(t)$.

[5] https://en.wikipedia.org/wiki/Michaelis-Menten_kinetics

To model how many people that get infected in a small time interval Δt, we reason as with reactions in Sect. 4.6. The possible number of pairings ("collisions") between susceptibles and infected is SI. A fraction of these, $\beta \Delta t SI$, will actually meet and the infected succeed in infecting the susceptible, where β is a parameter to be empirically estimated. This leads to a loss of susceptibles and a gain of infected:

$$S(t+\Delta t) = S(t) - \beta \Delta t SI,$$
$$I(t+\Delta t) = I(t) + \beta \Delta t SI\,.$$

In the same time interval, a fraction $\nu \Delta t I$ of the infected is recovered. It follows from Sect. 4.5.4 that the parameter ν^{-1} is interpreted as the average waiting time to leave the I category, i.e., the average length of the disease. The $\nu \Delta t I$ term is a loss for the I category, but a gain for the R category:

$$I(t+\Delta t) = I(t) + \beta \Delta t SI - \nu \Delta t I, \quad R(t+\Delta t) = R(t) + \nu \Delta t I\,.$$

Dividing these equations by Δt and going to the limit $\Delta t \rightarrow 0$, gives the ODE system

$$\frac{dS}{dt} = -\beta SI, \tag{4.27}$$

$$\frac{dI}{dt} = \beta SI - \nu I, \tag{4.28}$$

$$\frac{dR}{dt} = \nu I, \tag{4.29}$$

with initial values $S(0) = S_0$, $I(0) = I_0$, and $R(0) = 0$. By adding the equations, we realize that

$$\frac{dS}{dt} + \frac{dI}{dt} + \frac{dR}{dt} = 0 \quad \Rightarrow \quad S + I + R = \text{const} = N,$$

where N is the total number in the population under consideration. This property can be used as a partial verification during simulations.

Equations (4.27)–(4.29) are known as the SIR model in epidemiology. The model can easily be extended to incorporate vaccination programs, immunity loss after some time, etc. Typical diseases that can be simulated by the SIR model and its variants are measles, smallpox, flu, plague, and HIV.

4.8 Predator-Prey Models in Ecology

A model for the interaction of predator and prey species can be based on reasoning from population dynamics and the SIR model. Let $H(t)$ be the number of preys in a region, and let $L(t)$ be the number of predators. For example, H may be hares and L lynx, or rabbits and foxes.

The population of the prey evolves due to births and deaths, exactly as in a population dynamics model from Sect. 4.2.1. During a time interval Δt the increase in the population is therefore

$$H(t+\Delta t) - H(t) = a \Delta t H(t),$$

where a is a parameter to be measured from data. The increase is proportional to H, and the proportionality constant $a\Delta t$ is proportional to Δt, because doubling the interval will double the increase.

However, the prey population has an additional loss because they are eaten by predators. All the prey and predator animals can form LH pairs in total (assuming all individuals meet randomly). A small fraction $b\Delta t$ of such meetings, during a time interval Δt, ends up with the predator eating the prey. The increase in the prey population is therefore adjusted to

$$H(t+\Delta t) - H(t) = a\Delta t H(t) - b\Delta t H(t)L(t)\,.$$

The predator population increases as a result of eating preys. The amount of eaten preys is $b\Delta t LH$, but only a fraction $d\Delta t LH$ of this amount contributes to increasing the predator population. In addition, predators die and this loss is set to $c\Delta t L$. To summarize, the increase in the predator population is given by

$$L(t+\Delta t) - L(t) = d\Delta t L(t)H(t) - c\Delta t L(t)\,.$$

Dividing by Δt in the equations for H and L and letting $t \to 0$ results in

$$\lim_{\Delta t\to 0} \frac{H(t+\Delta t) - H(t)}{\Delta t} = H'(t) = aH(t) - bL(t)H(t),$$
$$\lim_{\Delta t\to 0} \frac{L(t+\Delta t) - L(t)}{\Delta t} = L'(t) = dL(t)H(t) - cL(t)\,.$$

We can simplify the notation to the following two ODEs:

$$H' = H(a - bL), \tag{4.30}$$
$$L' = L(dH - c)\,. \tag{4.31}$$

This is a so-called Lokta-Volterra model. It contains four parameters that must be estimated from data: a, b, c, and d. In addition, two initial conditions are needed for $H(0)$ and $L(0)$.

4.9 Decay of Atmospheric Pressure with Altitude

4.9.1 The General Model

Vertical equilibrium of air in the atmosphere is governed by the equation

$$\frac{dp}{dz} = -\varrho g\,. \tag{4.32}$$

Here, $p(z)$ is the air pressure, ϱ is the density of air, and $g = 9.807\,\mathrm{m/s^2}$ is a standard value of the acceleration of gravity. (Equation (4.32) follows directly from the general Navier-Stokes equations for fluid motion, with the assumption that the air does not move.)

The pressure is related to density and temperature through the ideal gas law

$$\varrho = \frac{Mp}{R^*T}, \tag{4.33}$$

where M is the molar mass of the Earth's air (0.029 kg/mol), R^* is the universal gas constant (8.314 Nm/(mol K)), and T is the temperature in Kelvin. All variables p, ϱ, and T vary with the height z. Inserting (4.33) in (4.32) results in an ODE with a variable coefficient:

$$\frac{dp}{dz} = -\frac{Mg}{R^*T(z)}p\,. \tag{4.34}$$

4.9.2 Multiple Atmospheric Layers

The atmosphere can be approximately modeled by seven layers. In each layer, (4.34) is applied with a linear temperature of the form

$$T(z) = \bar{T}_i + L_i(z - h_i),$$

where $z = h_i$ denotes the bottom of layer number i, having temperature $\bar{T}_i$, and L_i is a constant in layer number i. The table below lists h_i (m), $\bar{T}_i$ (K), and L_i (K/m) for the layers $i = 0, \ldots, 6$.

i	h_i	$\bar{T}_i$	L_i
0	0	288	−0.0065
1	11,000	216	0.0
2	20,000	216	0.001
3	32,000	228	0.0028
4	47,000	270	0.0
5	51,000	270	−0.0028
6	71,000	214	−0.002

For implementation it might be convenient to write (4.34) on the form

$$\frac{dp}{dz} = -\frac{Mg}{R^*(\bar{T}(z) + L(z)(z - h(z)))}p, \tag{4.35}$$

where $\bar{T}(z)$, $L(z)$, and $h(z)$ are piecewise constant functions with values given in the table. The value of the pressure at the sea level $z = 0$, $p_0 = p(0)$, is 101,325 Pa.

4.9.3 Simplifications

Constant layer temperature One common simplification is to assume that the temperature is constant within each layer. This means that $L = 0$.

One-layer model Another commonly used approximation is to work with one layer instead of seven. This one-layer model[6] is based on $T(z) = T_0 - Lz$,

[6] http://en.wikipedia.org/wiki/Density_of_air

with sea level standard temperature $T_0 = 288\,\text{K}$ and temperature lapse rate $L = 0.0065\,\text{K/m}$.

4.10 Compaction of Sediments

Sediments, originally made from materials like sand and mud, get compacted through geological time by the weight of new material that is deposited on the sea bottom. The porosity ϕ of the sediments tells how much void (fluid) space there is between the sand and mud grains. The porosity drops with depth, due to the weight of the sediments above. This makes the void space shrink, and thereby compaction increases.

A typical assumption is that the change in ϕ at some depth z is negatively proportional to ϕ. This assumption leads to the differential equation problem

$$\frac{d\phi}{dz} = -c\phi, \quad \phi(0) = \phi_0, \tag{4.36}$$

where the z axis points downwards, $z = 0$ is the surface with known porosity, and $c > 0$ is a constant.

The upper part of the Earth's crust consists of many geological layers stacked on top of each other, as indicated in Fig. 4.1. The model (4.36) can be applied for each layer. In layer number i, we have the unknown porosity function $\phi_i(z)$ fulfilling $\phi_i'(z) = -c_i z$, since the constant c in the model (4.36) depends on the type of sediment in the layer. Alternatively, we can use (4.36) to describe the porosity through all layers if c is taken as a piecewise constant function of z, equal to c_i in layer i. From the figure we see that new layers of sediments are deposited on top of older ones as time progresses. The compaction, as measured by ϕ, is rapid in the beginning and then decreases (exponentially) with depth in each layer.

When we drill a well at present time through the right-most column of sediments in Fig. 4.1, we can measure the thickness of the sediment in (say) the bottom layer.

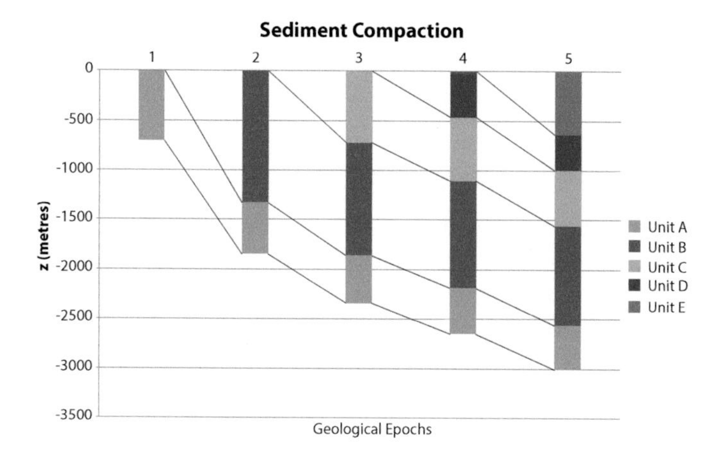

Fig. 4.1 Illustration of the compaction of geological layers (*with different colors*) through time

Let L_1 be this thickness. Assuming that the volume of sediment remains constant through time, we have that the initial volume, $\int_0^{L_{1,0}} \phi_1 dz$, must equal the volume seen today, $\int_{\ell-L_1}^{\ell} \phi_1 dz$, where ℓ is the depth of the bottom of the sediment in the present day configuration. After having solved for ϕ_1 as a function of z, we can then find the original thickness $L_{1,0}$ of the sediment from the equation

$$\int_0^{L_{1,0}} \phi_1 dz = \int_{\ell-L_1}^{\ell} \phi_1 dz\,.$$

In hydrocarbon exploration it is important to know $L_{1,0}$ and the compaction history of the various layers of sediments.

4.11 Vertical Motion of a Body in a Viscous Fluid

A body moving vertically through a fluid (liquid or gas) is subject to three different types of forces: the gravity force, the drag force[7], and the buoyancy force.

4.11.1 Overview of Forces

Taking the upward direction as positive, the gravity force is $F_g = -mg$, where m is the mass of the body and g is the acceleration of gravity. The uplift or buoyancy force ("Archimedes force") is $F_b = \varrho g V$, where ϱ is the density of the fluid and V is the volume of the body.

The drag force is of two types, depending on the Reynolds number

$$\text{Re} = \frac{\varrho d|v|}{\mu}, \tag{4.37}$$

where d is the diameter of the body in the direction perpendicular to the flow, v is the velocity of the body, and μ is the dynamic viscosity of the fluid. When $\text{Re} < 1$, the drag force is fairly well modeled by the so-called Stokes' drag, which for a spherical body of diameter d reads

$$F_d^{(S)} = -3\pi d\mu v\,. \tag{4.38}$$

Quantities are taken as positive in the upwards vertical direction, so if $v > 0$ and the body moves upwards, the drag force acts downwards and become negative, in accordance with the minus sign in expression for $F_d^{(S)}$.

For large Re, typically $\text{Re} > 10^3$, the drag force is quadratic in the velocity:

$$F_d^{(q)} = -\frac{1}{2} C_D \varrho A |v| v, \tag{4.39}$$

[7] http://en.wikipedia.org/wiki/Drag_(physics)

where C_D is a dimensionless drag coefficient depending on the body's shape, and A is the cross-sectional area as produced by a cut plane, perpendicular to the motion, through the thickest part of the body. The superscripts ^{q} and ^{S} in $F_d^{(S)}$ and $F_d^{(q)}$ indicate Stokes' drag and quadratic drag, respectively.

4.11.2 Equation of Motion

All the mentioned forces act in the vertical direction. Newton's second law of motion applied to the body says that the sum of these forces must equal the mass of the body times its acceleration a in the vertical direction.

$$ma = F_g + F_d^{(S)} + F_b .$$

Here we have chosen to model the fluid resistance by the Stokes' drag. Inserting the expressions for the forces yields

$$ma = -mg - 3\pi d\mu v + \varrho g V .$$

The unknowns here are v and a, i.e., we have two unknowns but only one equation. From kinematics in physics we know that the acceleration is the time derivative of the velocity: $a = dv/dt$. This is our second equation. We can easily eliminate a and get a single differential equation for v:

$$m\frac{dv}{dt} = -mg - 3\pi d\mu v + \varrho g V .$$

A small rewrite of this equation is handy: We express m as $\varrho_b V$, where ϱ_b is the density of the body, and we divide by the mass to get

$$v'(t) = -\frac{3\pi d\mu}{\varrho_b V}v + g\left(\frac{\varrho}{\varrho_b} - 1\right) . \tag{4.40}$$

We may introduce the constants

$$a = \frac{3\pi d\mu}{\varrho_b V}, \quad b = g\left(\frac{\varrho}{\varrho_b} - 1\right), \tag{4.41}$$

so that the structure of the differential equation becomes obvious:

$$v'(t) = -av(t) + b . \tag{4.42}$$

The corresponding initial condition is $v(0) = v_0$ for some prescribed starting velocity v_0.

This derivation can be repeated with the quadratic drag force $F_d^{(q)}$, leading to the result

$$v'(t) = -\frac{1}{2}C_D\frac{\varrho A}{\varrho_b V}|v|v + g\left(\frac{\varrho}{\varrho_b} - 1\right) . \tag{4.43}$$

Defining

$$a = \frac{1}{2} C_D \frac{\varrho A}{\varrho_b V}, \qquad (4.44)$$

and b as above, we can write (4.43) as

$$v'(t) = -a|v|v + b\,. \qquad (4.45)$$

4.11.3 Terminal Velocity

An interesting aspect of (4.42) and (4.45) is whether v will approach a final constant value, the so-called *terminal velocity* v_T, as $t \to \infty$. A constant v means that $v'(t) \to 0$ as $t \to \infty$ and therefore the terminal velocity v_T solves

$$0 = -av_T + b$$

and

$$0 = -a|v_T|v_T + b\,.$$

The former equation implies $v_T = b/a$, while the latter has solutions $v_T = -\sqrt{|b|/a}$ for a falling body ($v_T < 0$) and $v_T = \sqrt{b/a}$ for a rising body ($v_T > 0$).

4.11.4 A Crank–Nicolson Scheme

Both governing equations, the Stokes' drag model (4.42) and the quadratic drag model (4.45), can be readily solved by the Forward Euler scheme. For higher accuracy one can use the Crank–Nicolson method, but a straightforward application of this method gives a nonlinear equation in the new unknown value v^{n+1} when applied to (4.45):

$$\frac{v^{n+1} - v^n}{\Delta t} = -a\frac{1}{2}(|v^{n+1}|v^{n+1} + |v^n|v^n) + b\,. \qquad (4.46)$$

The first term on the right-hand side of (4.46) is the arithmetic average of $-|v|v$ evaluated at time levels n and $n+1$.

Instead of approximating the term $-|v|v$ by an arithmetic average, we can use a *geometric mean*:

$$(|v|v)^{n+\frac{1}{2}} \approx |v^n|v^{n+1}\,. \qquad (4.47)$$

The error is of second order in Δt, just as for the arithmetic average and the centered finite difference approximation in (4.46). With the geometric mean, the resulting discrete equation

$$\frac{v^{n+1} - v^n}{\Delta t} = -a|v^n|v^{n+1} + b$$

becomes a *linear* equation in v^{n+1}, and we can therefore easily solve for v^{n+1}:

$$v^{n+1} = \frac{v_n + \Delta t b^{n+\frac{1}{2}}}{1 + \Delta t a^{n+\frac{1}{2}} |v^n|} \,. \tag{4.48}$$

Using a geometric mean instead of an arithmetic mean in the Crank–Nicolson scheme is an attractive method for avoiding a nonlinear algebraic equation when discretizing a nonlinear ODE.

4.11.5 Physical Data

Suitable values of μ are $1.8 \cdot 10^{-5}$ Pa s for air and $8.9 \cdot 10^{-4}$ Pa s for water. Densities can be taken as $1.2\,\mathrm{kg/m^3}$ for air and as $1.0 \cdot 10^3\,\mathrm{kg/m^3}$ for water. For considerable vertical displacement in the atmosphere one should take into account that the density of air varies with the altitude, see Sect. 4.9. One possible density variation arises from the one-layer model in the mentioned section.

Any density variation makes b time dependent and we need $b^{n+\frac{1}{2}}$ in (4.48). To compute the density that enters $b^{n+\frac{1}{2}}$ we must also compute the vertical position $z(t)$ of the body. Since $v = dz/dt$, we can use a centered difference approximation:

$$\frac{z^{n+\frac{1}{2}} - z^{n-\frac{1}{2}}}{\Delta t} = v^n \quad \Rightarrow \quad z^{n+\frac{1}{2}} = z^{n-\frac{1}{2}} + \Delta t\, v^n \,.$$

This $z^{n+\frac{1}{2}}$ is used in the expression for b to compute $\varrho(z^{n+\frac{1}{2}})$ and then $b^{n+\frac{1}{2}}$.

The drag coefficient[8] C_D depends heavily on the shape of the body. Some values are: 0.45 for a sphere, 0.42 for a semi-sphere, 1.05 for a cube, 0.82 for a long cylinder (when the center axis is in the vertical direction), 0.75 for a rocket, 1.0-1.3 for a man in upright position, 1.3 for a flat plate perpendicular to the flow, and 0.04 for a streamlined, droplet-like body.

4.11.6 Verification

To verify the program, one may assume a heavy body in air such that the F_b force can be neglected, and further assume a small velocity such that the air resistance F_d can also be neglected. This can be obtained by setting μ and ϱ to zero. The motion then leads to the velocity $v(t) = v_0 - gt$, which is linear in t and therefore should be reproduced to machine precision (say tolerance 10^{-15}) by any implementation based on the Crank–Nicolson or Forward Euler schemes.

Another verification, but not as powerful as the one above, can be based on computing the terminal velocity and comparing with the exact expressions. The advantage of this verification is that we can also test the situation $\varrho \neq 0$.

As always, the method of manufactured solutions can be applied to test the implementation of all terms in the governing equation, but then the solution has no physical relevance in general.

[8] http://en.wikipedia.org/wiki/Drag_coefficient

4.11.7 Scaling

Applying scaling, as described in Sect. 4.1, will for the linear case reduce the need to estimate values for seven parameters down to choosing one value of a single dimensionless parameter

$$\beta = \frac{\varrho_b g V\left(\frac{\varrho}{\varrho_b} - 1\right)}{3\pi d\mu I},$$

provided $I \neq 0$. If the motion starts from rest, $I = 0$, the scaled problem reads

$$\bar{u}' = 1 - \bar{u}, \quad \bar{u}(0) = 0,$$

and there is no need for estimating physical parameters (!). This means that there is a single universal solution to the problem of a falling body starting from rest: $\bar{u}(t) = 1 - e^{-\bar{t}}$. All real physical cases correspond to stretching the $\bar{t}$ axis and the $\bar{u}$ axis in this dimensionless solution. More precisely, the physical velocity $u(t)$ is related to the dimensionless velocity $\bar{u}(\bar{t})$ through

$$u = \frac{\varrho_b g V\left(\frac{\varrho}{\varrho_b} - 1\right)}{3\pi d\mu}\bar{u}(t/(g(\varrho/\varrho_b - 1))) = \frac{\varrho_b g V\left(\frac{\varrho}{\varrho_b} - 1\right)}{3\pi d\mu}(1 - e^{t/(g(\varrho/\varrho_b - 1))}).$$

4.12 Viscoelastic Materials

When stretching a rod made of a perfectly elastic material, the elongation (change in the rod's length) is proportional to the magnitude of the applied force. Mathematical models for material behavior under application of external forces use *strain* ε and *stress* σ instead of elongation and forces. Strain is relative change in elongation and stress is force per unit area. An elastic material has a linear relation between stress and strain: $\sigma = E\varepsilon$. This is a good model for many materials, but frequently the velocity of the deformation (or more precisely, the strain rate ε') also influences the stress. This is particularly the case for materials like organic polymers, rubber, and wood. When the stress depends on both the strain and the strain rate, the material is said to be viscoelastic. A common model relating forces to deformation is the Kelvin–Voigt model[9]:

$$\sigma(t) = E\varepsilon(t) + \eta\varepsilon'(t). \tag{4.49}$$

Compared to a perfectly elastic material, which deforms instantaneously when a force is acting, a Kelvin–Voigt material will spend some time to elongate. For example, when an elastic rod is subject to a constant force $\boldsymbol{\sigma}$ at $t = 0$, the strain immediately adjusts to $\varepsilon = \sigma/E$. A Kelvin–Voigt material, however, has a response $\varepsilon(t) = \frac{\sigma}{E}(1 - e^{Et/\eta})$. Removing the force when the strain is $\varepsilon(t_1) = I$ will for an elastic material immediately bring the strain back to zero, while a Kelvin–Voigt material will decay: $\varepsilon = Ie^{-(t-t_1)E/\eta)}$.

[9] https://en.wikipedia.org/wiki/Kelvin-Voigt_material

Introducing u for ε and treating $\sigma(t)$ as a given function, we can write the Kelvin–Voigt model in our standard form

$$u'(t) = -au(t) + b(t), \tag{4.50}$$

with $a = E/\eta$ and $b(t) = \sigma(t)/\eta$. An initial condition, usually $u(0) = 0$, is needed.

4.13 Decay ODEs from Solving a PDE by Fourier Expansions

Suppose we have a partial differential equation

$$\frac{\partial u}{\partial t} = \alpha\frac{\partial^2 u}{\partial x^2} + f(x,t),$$

with boundary conditions $u(0,t) = u(L,t) = 0$ and initial condition $u(x,0) = I(x)$. One may express the solution as

$$u(x,t) = \sum_{k=1}^{m} A_k(t)e^{ikx\pi/L},$$

for appropriate unknown functions A_k, $k = 1,\ldots,m$. We use the complex exponential $e^{ikx\pi/L}$ for easy algebra, but the physical u is taken as the real part of any complex expression. Note that the expansion in terms of $e^{ikx\pi/L}$ is compatible with the boundary conditions: all functions $e^{ikx\pi/L}$ vanish for $x = 0$ and $x = L$. Suppose we can express $I(x)$ as

$$I(x) = \sum_{k=1}^{m} I_k e^{ikx\pi/L}.$$

Such an expansion can be computed by well-known Fourier expansion techniques, but those details are not important here. Also, suppose we can express the given $f(x,t)$ as

$$f(x,t) = \sum_{k=1}^{m} b_k(t)e^{ikx\pi/L}.$$

Inserting the expansions for u and f in the differential equations demands that all terms corresponding to a given k must be equal. The calculations result in the follow system of ODEs:

$$A_k'(t) = -\alpha\frac{k^2\pi^2}{L^2} + b_k(t), \quad k = 1,\ldots,m.$$

From the initial condition

$$u(x,0) = \sum_k A_k(0)e^{ikx\pi/L} = I(x) = \sum_k I_k e^{(ikx\pi/L)},$$

so it follows that $A_k(0) = I_k$, $k = 1,\ldots,m$. We then have m equations of the form $A_k' = -aA_k + b$, $A_k(0) = I_k$, for appropriate definitions of a and b. These ODE problems are independent of each other such that we can solve one problem at a time. The outlined technique is a quite common solution approach to partial differential equations.

Remark Since a_k depends on k and the stability of the Forward Euler scheme demands $a_k\Delta t \leq 1$, we get that $\Delta t \leq \alpha^{-1}L^2\pi^{-2}k^{-2}$ for this scheme. Usually, quite large k values are needed to accurately represent the given functions I and f so that Δt in the Forward Euler scheme needs to be very small for these large values of k. Therefore, the Crank–Nicolson and Backward Euler schemes, which allow larger Δt without any growth in the solutions, are more popular choices when creating time-stepping algorithms for partial differential equations of the type considered in this example.

4.14 Exercises

Problem 4.1: Radioactive decay of Carbon-14

The Carbon-14[10] isotope, whose radioactive decay is used extensively in dating organic material that is tens of thousands of years old, has a half-life of 5,730 years. Determine the age of an organic material that contains 8.4 % of its initial amount of Carbon-14. Use a time unit of 1 year in the computations. The uncertainty in the half time of Carbon-14 is ±40 years. What is the corresponding uncertainty in the estimate of the age?

Hint 1 Let A be the amount of Carbon-14. The ODE problem is then $A'(t) = -aA(t)$, $A(0) = I$. Introduced the scaled amount $u = A/I$. The ODE problem for u is $u' = -au$, $u(0) = 1$. Measure time in years. Simulate until the first mesh point t_m such that $u(t_m) \leq 0.084$.

Hint 2 Use simulations with $5{,}730 \pm 40$ y as input and find the corresponding uncertainty interval for the result.

Filename: `carbon14`.

Exercise 4.2: Derive schemes for Newton's law of cooling

Show in detail how we can apply the ideas of the Forward Euler, Backward Euler, and Crank–Nicolson discretizations to derive explicit computational formulas for new temperature values in Newton's law of cooling (see Sect. 4.4):

$$\frac{dT}{dt} = -k(T - T_s(t)), \quad T(0) = T_0\,.$$

Here, T is the temperature of the body, $T_s(t)$ is the temperature of the surroundings, t is time, k is the heat transfer coefficient, and T_0 is the initial temperature of the body. Summarize the discretizations in a θ-rule such that you can get the three schemes from a single formula by varying the θ parameter.

Filename: `schemes_cooling`.

[10] http://en.wikipedia.org/wiki/Carbon-14

Exercise 4.3: Implement schemes for Newton's law of cooling
The goal of this exercise is to implement the schemes from Exercise 4.2 and investigate several approaches for verifying the implementation.

a) Implement the θ-rule from Exercise 4.2 in a function

```
cooling(T0, k, T_s, t_end, dt, theta=0.5)
```

where `T0` is the initial temperature, `k` is the heat transfer coefficient, `T_s` is a function of `t` for the temperature of the surroundings, `t_end` is the end time of the simulation, `dt` is the time step, and `theta` corresponds to θ. The `cooling` function should return the temperature as an array `T` of values at the mesh points and the time mesh `t`.

b) In the case $\lim_{t\to\infty} T_s(t) = C = \text{const}$, explain why $T(t) \to C$. Construct an example where you can illustrate this property in a plot. Implement a corresponding test function that checks the correctness of the asymptotic value of the solution.

c) A piecewise constant surrounding temperature,

$$T_s(t) = \begin{cases} C_0, & 0 \le t \le t^* \\ C_1, & t > t^*, \end{cases}$$

corresponds to a sudden change in the environment at $t = t^*$. Choose $C_0 = 2T_0$, $C_1 = \frac{1}{2}T_0$, and $t^* = 3/k$. Plot the solution $T(t)$ and explain why it seems physically reasonable.

d) We know from the ODE $u' = -au$ that the Crank–Nicolson scheme can give non-physical oscillations for $\Delta t > 2/a$. In the present problem, this results indicates that the Crank–Nicolson scheme give undesired oscillations for $\Delta t > 2/k$. Discuss if this a potential problem in the physical case from c).

e) Find an expression for the exact solution of $T' = -k(T - T_s(t))$, $T(0) = T_0$. Construct a test case and compare the numerical and exact solution in a plot. Find a value of the time step Δt such that the two solution curves cannot (visually) be distinguished from each other. Many scientists will claim that such a plot provides evidence for a correct implementation, but point out why there still may be errors in the code. Can you introduce bugs in the `cooling` function and still achieve visually coinciding curves?

Hint The exact solution can be derived by multiplying (4.8) by the integrating factor e^{kt}.

f) Implement a test function for checking that the solution returned by the `cooling` function is identical to the exact numerical solution of the problem (to machine precision) when T_s is constant.

Hint The exact solution of the discrete equations in the case T_s is a constant can be found by introducing $u = T - T_s$ to get a problem $u' = -ku$, $u(0) = T_0 - T_s$. The solution of the discrete equations is then of the form $u^n = (T_0 - T_s)A^n$ for some amplification factor A. The expression for T^n is then $T^n = T_s(t_n) + u^n = T_s + (T_0 - T_s)A^n$. We find that

$$A = \frac{1 - (1-\theta)k\Delta t}{1 + \theta k \Delta t}.$$

The test function, testing several θ values for a quite coarse mesh, may take the form

```
def test_discrete_solution():
    """
    Compare the numerical solution with an exact solution
    of the scheme when the T_s is constant.
    """
    T_s = 10
    T0 = 2
    k = 1.2
    dt = 0.1     # can use any mesh
    N_t = 6      # any no of steps will do
    t_end = dt*N_t
    t = np.linspace(0, t_end, N_t+1)

    for theta in [0, 0.5, 1, 0.2]:
        u, t = cooling(T0, k, lambda t: T_s , t_end, dt, theta)
        A = (1 - (1-theta)*k*dt)/(1 + theta*k*dt)
        u_discrete_exact = T_s + (T0-T_s)*A**(np.arange(len(t)))
        diff = np.abs(u - u_discrete_exact).max()
        print 'diff computed and exact discrete solution:', diff
        tol = 1E-14
        success = diff < tol
        assert success, 'diff=%g' % diff
```

Running this function shows that the `diff` variable is `3.55E-15` as maximum so a tolerance of 10^{-14} is appropriate. This is a good test that the `cooling` function works!

Filename: `cooling`.

Exercise 4.4: Find time of murder from body temperature

A detective measures the temperature of a dead body to be 26.7 °C at 2 pm. One hour later the temperature is 25.8 °C. The question is when death occurred.

Assume that Newton's law of cooling (4.8) is an appropriate mathematical model for the evolution of the temperature in the body. First, determine k in (4.8) by formulating a Forward Euler approximation with one time steep from time 2 am to time 3 am, where knowing the two temperatures allows for finding k. Assume the temperature in the air to be 20 °C. Thereafter, simulate the temperature evolution from the time of murder, taken as $t = 0$, when $T = 37$ °C, until the temperature reaches 25.8 °C. The corresponding time allows for answering when death occurred.

Filename: `detective`.

Exercise 4.5: Simulate an oscillating cooling process

The surrounding temperature T_s in Newton's law of cooling (4.8) may vary in time. Assume that the variations are periodic with period P and amplitude a around a constant mean temperature T_m:

$$T_s(t) = T_m + a \sin\left(\frac{2\pi}{P}t\right) . \tag{4.51}$$

Simulate a process with the following data: $k = 0.05\,\text{min}^{-1}$, $T(0) = 5\,°\text{C}$, $T_m = 25\,°\text{C}$, $a = 2.5\,°\text{C}$, and $P = 1\,\text{h}$, $P = 10\,\text{min}$, and $P = 6\,\text{h}$. Plot the T solutions and T_s in the same plot.

Filename: `osc_cooling`.

Exercise 4.6: Simulate stochastic radioactive decay
The purpose of this exercise is to implement the stochastic model described in Sect. 4.5 and show that its mean behavior approximates the solution of the corresponding ODE model.

The simulation goes on for a time interval $[0, T]$ divided into N_t intervals of length Δt. We start with N_0 atoms. In some time interval, we have N atoms that have survived. Simulate N Bernoulli trials with probability $\lambda\Delta t$ in this interval by drawing N random numbers, each being 0 (survival) or 1 (decay), where the probability of getting 1 is $\lambda\Delta t$. We are interested in the number of decays, d, and the number of survived atoms in the next interval is then $N - d$. The Bernoulli trials are simulated by drawing N uniformly distributed real numbers on $[0, 1]$ and saying that 1 corresponds to a value less than $\lambda\Delta t$:

```
# Given lambda_, dt, N
import numpy as np
uniform = np.random.uniform(N)
Bernoulli_trials = np.asarray(uniform < lambda_*dt, dtype=np.int)
d = Bernoulli_trials.size
```

Observe that `uniform < lambda_*dt` is a boolean array whose true and false values become 1 and 0, respectively, when converted to an integer array.

Repeat the simulation over $[0, T]$ a large number of times, compute the average value of N in each interval, and compare with the solution of the corresponding ODE model.

Filename: `stochastic_decay`.

Problem 4.7: Radioactive decay of two substances
Consider two radioactive substances A and B. The nuclei in substance A decay to form nuclei of type B with a half-life $A_{1/2}$, while substance B decay to form type A nuclei with a half-life $B_{1/2}$. Letting u_A and u_B be the fractions of the initial amount of material in substance A and B, respectively, the following system of ODEs governs the evolution of $u_A(t)$ and $u_B(t)$:

$$\frac{1}{\ln 2}u_A' = u_B/B_{1/2} - u_A/A_{1/2}, \tag{4.52}$$

$$\frac{1}{\ln 2}u_B' = u_A/A_{1/2} - u_B/B_{1/2}, \tag{4.53}$$

with $u_A(0) = u_B(0) = 1$.

a) Make a simulation program that solves for $u_A(t)$ and $u_B(t)$.
b) Verify the implementation by computing analytically the limiting values of u_A and u_B as $t \to \infty$ (assume $u'_A, u'_B \to 0$) and comparing these with those obtained numerically.
c) Run the program for the case of $A_{1/2} = 10$ minutes and $B_{1/2} = 50$ minutes. Use a time unit of 1 minute. Plot u_A and u_B versus time in the same plot.

Filename: radioactive_decay_2subst.

Exercise 4.8: Simulate a simple chemical reaction
Consider the simple chemical reaction where a substance A is turned into a substance B according to

$$\frac{d[A]}{dt} = -k[A],$$
$$\frac{d[B]}{dt} = k[A],$$

where $[A]$ and $[B]$ are the concentrations of A and B, respectively. It may be a challenge to find appropriate values of k, but we can avoid this problem by working with a scaled model (as explained in Sect. 4.1). Scale the model above, using a time scale $1/k$, and use the initial concentration of $[A]$ as scale for $[A]$ and $[B]$. Show that the scaled system reads

$$\frac{du}{dt} = -u,$$
$$\frac{dv}{dt} = u,$$

with initial conditions $u(0) = 1$, and $v(0) = \alpha$, where $\alpha = [B](0)/[A](0)$ is a dimensionless number, and u and v are the scaled concentrations of $[A]$ and $[B]$, respectively. Implement a numerical scheme that can be used to find the solutions $u(t)$ and $v(t)$. Visualize u and v in the same plot.
Filename: chemcial_kinetics_AB.

Exercise 4.9: Simulate an *n*-th order chemical reaction
An n-order chemical reaction, generalizing the model in Exercise 4.8, takes the form

$$\frac{d[A]}{dt} = -k[A]^n,$$
$$\frac{d[B]}{dt} = k[A]^n,$$

where symbols are as defined in Exercise 4.8. Bring this model on dimensionless form, using a time scale $[A](0)^{n-1}/k$, and show that the dimensionless model simplifies to

$$\frac{du}{dt} = -u^n,$$
$$\frac{dv}{dt} = u^n,$$

with $u(0) = 1$ and $v(0) = \alpha = [B](0)/[A](0)$. Solve numerically for $u(t)$ and show a plot with u for $n = 0.5, 1, 2, 4$.

Filename: `chemcial_kinetics_ABn`.

Exercise 4.10: Simulate a biochemical process

The purpose of this exercise is to simulate the ODE system (4.23)–(4.26) modeling a simple biochemical process.

a) Scale (4.23)–(4.26) such that we can work with dimensionless parameters, which are easier to prescribe. Introduce

$$\bar{Q} = \frac{[ES]}{Q_c}, \quad \bar{P} = \frac{P}{P_c}, \quad \bar{S} = \frac{S}{S_0}, \quad \bar{E} = \frac{E}{E_0}, \quad \bar{t} = \frac{t}{t_c},$$

where appropriate scales are

$$Q_c = \frac{S_0 E_0}{K}, \quad P_c = Q_c, \quad t_c = \frac{1}{k_+ E_0},$$

with $K = (k_v + k_-)/k_+$ (Michaelis constant). Show that the scaled system becomes

$$\frac{d\bar{Q}}{d\bar{t}} = \alpha(\bar{E}\bar{S} - \bar{Q}), \tag{4.54}$$
$$\frac{d\bar{P}}{d\bar{t}} = \beta\bar{Q}, \tag{4.55}$$
$$\frac{d\bar{S}}{d\bar{t}} = -\bar{E}\bar{S} + (1 - \beta\alpha^{-1})\bar{Q}, \tag{4.56}$$
$$\epsilon\frac{d\bar{E}}{d\bar{t}} = -\bar{E}\bar{S} + \bar{Q}, \tag{4.57}$$

where we have three dimensionless parameters

$$\alpha = \frac{K}{E_0}, \quad \beta = \frac{k_v}{k_+ E_0}, \quad \epsilon = \frac{E_0}{S_0}.$$

The corresponding initial conditions are $\bar{Q} = \bar{P} = 0$ and $\bar{S} = \bar{E} = 1$.

b) Implement a function for solving (4.54)–(4.57).

c) There are two conservation equations implied by (4.23)–(4.26):

$$[ES] + [E] = E_0, \tag{4.58}$$
$$[ES] + [S] + [P] = S_0. \tag{4.59}$$

Derive these two equations. Use these properties in the function in b) to do a partial verification of the solution at each time step.

d) Simulate a case with $T = 8$, $\alpha = 1.5$ and $\beta = 1$, and two ϵ values: 0.9 and 0.1.

Filename: `biochem`.

Exercise 4.11: Simulate spreading of a disease

The SIR model (4.27)–(4.29) can be used to simulate spreading of an epidemic disease.

a) Estimating the parameter β is difficult so it can be handy to scale the equations. Use $t_c = 1/\nu$ as time scale, and scale S, I, and R by the population size $N = S(0)+I(0)+R(0)$. Show that the resulting dimensionless model becomes

$$\frac{d\bar{S}}{d\bar{t}} = -R_0\bar{S}\bar{I}, \tag{4.60}$$

$$\frac{d\bar{I}}{d\bar{t}} = R_0\bar{S}\bar{I} - \bar{I}, \tag{4.61}$$

$$\frac{d\bar{R}}{d\bar{t}} = I, \tag{4.62}$$

$$\bar{S}(0) = 1 - \alpha, \tag{4.63}$$

$$\bar{I}(0) = \alpha, \tag{4.64}$$

$$\bar{R}(0) = 0, \tag{4.65}$$

where R_0 and α are the only parameters in the problem:

$$R_0 = \frac{N\beta}{\nu}, \quad \alpha = \frac{I(0)}{N}.$$

A quantity with a bar denotes a dimensionless version of that quantity, e.g, $\bar{t}$ is dimensionless time, and $\bar{t} = \nu t$.

b) Show that the R_0 parameter governs whether the disease will spread or not at $t = 0$.

Hint Spreading means $dI/dt > 0$.

c) Implement the scaled SIR model. Check at every time step, as a verification, that $\bar{S} + \bar{I} + \bar{R} = 1$.

d) Simulate the spreading of a disease where $R_0 = 2, 5$ and 2 % of the population is infected at time $t = 0$.

Filename: `SIR`.

Exercise 4.12: Simulate predator-prey interaction

Section 4.8 describes a model for the interaction of predator and prey populations, such as lynx and hares.

a) Scale the equations (4.30)–(4.31). Use the initial population $H(0) = H_0$ of H has scale for H and L, and let the time scale be $1/(bH_0)$.
b) Implement the scaled model from a). Run illustrating cases how the two populations develop.
c) The scaling in a) used a scale for H and L based on the initial condition $H(0) = H_0$. An alternative scaling is to make the ODEs as simple as possible by introducing separate scales H_c and L_c for H and L, respectively. Fit H_c, L_c, and the time scale t_c such that there are as few dimensionless parameters as possible in the ODEs. Scale the initial conditions. Compare the number and type of dimensionless parameters with a).
d) Compute with the scaled model from c) and create plots to illustrate the typical solutions.

Filename: `predator_prey`.

Exercise 4.13: Simulate the pressure drop in the atmosphere
We consider the models for atmospheric pressure in Sect. 4.9. Make a program with three functions,

- one computing the pressure $p(z)$ using a seven-layer model and varying L,
- one computing $p(z)$ using a seven-layer model, but with constant temperature in each layer, and
- one computing $p(z)$ based on the one-layer model.

How can these implementations be verified? Should ease of verification impact how you code the functions? Compare the three models in a plot.

Filename: `atmospheric_pressure`.

Exercise 4.14: Make a program for vertical motion in a fluid
Implement the Stokes' drag model (4.40) and the quadratic drag model (4.43) from Sect. 4.11, using the Crank–Nicolson scheme and a geometric mean for $|v|v$ as explained, and assume constant fluid density. At each time level, compute the Reynolds number Re and choose the Stokes' drag model if Re < 1 and the quadratic drag model otherwise.

The computation of the numerical solution should take place either in a stand-alone function or in a solver class that looks up a problem class for physical data. Create a module and equip it with pytest/nose compatible test functions for automatically verifying the code.

Verification tests can be based on

- the terminal velocity (see Sect. 4.11),
- the exact solution when the drag force is neglected (see Sect. 4.11),
- the method of manufactured solutions (see Sect. 3.1.5) combined with computing convergence rates (see Sect. 3.1.6).

Use, e.g., a quadratic polynomial for the velocity in the method of manufactured solutions. The expected error is $\mathcal{O}(\Delta t^2)$ from the centered finite difference approximation and the geometric mean approximation for $|v|v$.

A solution that is linear in t will also be an exact solution of the discrete equations in many problems. Show that this is true for linear drag (by adding a source term that depends on t), but not for quadratic drag because of the geometric mean approximation. Use the method of manufactured solutions to add a source term *in the discrete equations for quadratic drag* such that a linear function of t is a solution. Add a test function for checking that the linear function is reproduced to machine precision in the case of both linear and quadratic drag.

Apply the software to a case where a ball rises in water. The buoyancy force is here the driving force, but the drag will be significant and balance the other forces after a short time. A soccer ball has radius 11 cm and mass 0.43 kg. Start the motion from rest, set the density of water, ϱ, to $1000\,\mathrm{kg/m^3}$, set the dynamic viscosity, μ, to 10^{-3} Pa s, and use a drag coefficient for a sphere: 0.45. Plot the velocity of the rising ball.

Filename: `vertical_motion`.

Project 4.15: Simulate parachuting

The aim of this project is to develop a general solver for the vertical motion of a body with quadratic air drag, verify the solver, apply the solver to a skydiver in free fall, and finally apply the solver to a complete parachute jump.

All the pieces of software implemented in this project should be realized as Python functions and/or classes and collected in one module.

a) Set up the differential equation problem that governs the velocity of the motion. The parachute jumper is subject to the gravity force and a quadratic drag force. Assume constant density. Add an extra source term to be used for program verification. Identify the input data to the problem.
b) Make a Python module for computing the velocity of the motion. Also equip the module with functionality for plotting the velocity.

Hint 1 Use the Crank–Nicolson scheme with a geometric mean of $|v|v$ in time to linearize the equation of motion with quadratic drag.

Hint 2 You can either use functions or classes for implementation. If you choose functions, make a function `solver` that takes all the input data in the problem as arguments and that returns the velocity (as a mesh function) and the time mesh. In case of a class-based implementation, introduce a problem class with the physical data and a solver class with the numerical data and a `solve` method that stores the velocity and the mesh in the class.

Allow for a time-dependent area and drag coefficient in the formula for the drag force.

c) Show that a linear function of t does not fulfill the discrete equations because of the geometric mean approximation used for the quadratic drag term. Fit a source term, as in the method of manufactured solutions, such that a linear function of t is a solution of the discrete equations. Make a test function to check that this solution is reproduced to machine precision.

d) The expected error in this problem goes like Δt^2 because we use a centered finite difference approximation with error $\mathcal{O}(\Delta t^2)$ and a geometric mean approximation with error $\mathcal{O}(\Delta t^2)$. Use the method of manufactured solutions combined with computing convergence rate to verify the code. Make a test function for checking that the convergence rate is correct.
e) Compute the drag force, the gravity force, and the buoyancy force as a function of time. Create a plot with these three forces.

Hint You can either make a function `forces(v, t, plot=None)` that returns the forces (as mesh functions) and `t`, and shows a plot on the screen and also saves the plot to a file with name stored in `plot` if `plot` is not `None`, or you can extend the solver class with computation of forces and include plotting of forces in the visualization class.

f) Compute the velocity of a skydiver in free fall before the parachute opens.

Hint Meade and Struthers [11] provide some data relevant to skydiving[11]. The mass of the human body and equipment can be set to 100 kg. A skydiver in spread-eagle formation has a cross-section of 0.5 m^2 in the horizontal plane. The density of air decreases with altitude, but can be taken as constant, 1 kg/m^3, for altitudes relevant to skydiving (0–4000 m). The drag coefficient for a man in upright position can be set to 1.2. Start with a zero velocity. A free fall typically has a terminating velocity of 45 m/s. (This value can be used to tune other parameters.)

g) The next task is to simulate a parachute jumper during free fall and after the parachute opens. At time t_p, the parachute opens and the drag coefficient and the cross-sectional area change dramatically. Use the program to simulate a jump from $z = 3000$ m to the ground $z = 0$. What is the maximum acceleration, measured in units of g, experienced by the jumper?

Hint Following Meade and Struthers [11], one can set the cross-section area perpendicular to the motion to 44 m^2 when the parachute is open. Assume that it takes 8 s to increase the area linearly from the original to the final value. The drag coefficient for an open parachute can be taken as 1.8, but tuned using the known value of the typical terminating velocity reached before landing: 5.3 m/s. One can take the drag coefficient as a piecewise constant function with an abrupt change at t_p. The parachute is typically released after $t_p = 60$ s, but larger values of t_p can be used to make plots more illustrative.
Filename: `parachuting`.

Exercise 4.16: Formulate vertical motion in the atmosphere
Vertical motion of a body in the atmosphere needs to take into account a varying air density if the range of altitudes is many kilometers. In this case, ϱ varies with the altitude z. The equation of motion for the body is given in Sect. 4.11. Let us assume quadratic drag force (otherwise the body has to be very, very small). A differential

[11] http://en.wikipedia.org/wiki/Parachuting

equation problem for the air density, based on the information for the one-layer atmospheric model in Sect. 4.9, can be set up as

$$p'(z) = -\frac{Mg}{R^*(T_0 + Lz)}p, \tag{4.66}$$

$$\varrho = p\frac{M}{R^*T}. \tag{4.67}$$

To evaluate $p(z)$ we need the altitude z. From the principle that the velocity is the derivative of the position we have that

$$z'(t) = v(t), \tag{4.68}$$

where v is the velocity of the body.

Explain in detail how the governing equations can be discretized by the Forward Euler and the Crank–Nicolson methods. Discuss pros and cons of the two methods.

Filename: `falling_in_variable_density`.

Exercise 4.17: Simulate vertical motion in the atmosphere
Implement the Forward Euler or the Crank–Nicolson scheme derived in Exercise 4.16. Demonstrate the effect of air density variation on a falling human, e.g., the famous fall of Felix Baumgartner[12]. The drag coefficient can be set to 1.2.

Filename: `falling_in_variable_density`.

Problem 4.18: Compute $y = |x|$ by solving an ODE
Consider the ODE problem

$$y'(x) = \begin{cases} -1, & x < 0, \\ 1, & x \geq 0 \end{cases} \quad x \in (-1, 1], \quad y(1-) = 1,$$

which has the solution $y(x) = |x|$. Using a mesh $x_0 = -1$, $x_1 = 0$, and $x_2 = 1$, calculate by hand y_1 and y_2 from the Forward Euler, Backward Euler, Crank–Nicolson, and Leapfrog methods. Use all of the former three methods for computing the y_1 value to be used in the Leapfrog calculation of y_2. Thereafter, visualize how these schemes perform for a uniformly partitioned mesh with $N = 10$ and $N = 11$ points.

Filename: `signum`.

Problem 4.19: Simulate fortune growth with random interest rate
The goal of this exercise is to compute the value of a fortune subject to inflation and a random interest rate. Suppose that the inflation is constant at i percent per year and that the annual interest rate, p, changes randomly at each time step, starting at some value p_0 at $t = 0$. The random change is from a value p^n at $t = t_n$ to $p_n + \Delta p$ with probability 0.25 and $p_n - \Delta p$ with probability 0.25. No change occurs with probability 0.5. There is also no change if p^{n+1} exceeds 15 or becomes below 1. Use a time step of one month, $p_0 = i$, initial fortune scaled to 1, and simulate 1000

[12] http://en.wikipedia.org/wiki/Felix_Baumgartner

scenarios of length 20 years. Compute the mean evolution of one unit of money and the corresponding standard deviation. Plot the mean curve along with the mean plus one standard deviation and the mean minus one standard deviation. This will illustrate the uncertainty in the mean curve.

Hint 1 The following code snippet computes p^{n+1}:

```
import random

def new_interest_rate(p_n, dp=0.5):
    r = random.random()  # uniformly distr. random number in [0,1)
    if 0 <= r < 0.25:
        p_np1 = p_n + dp
    elif 0.25 <= r < 0.5:
        p_np1 = p_n - dp
    else:
        p_np1 = p_n
    return (p_np1 if 1 <= p_np1 <= 15 else p_n)
```

Hint 2 If $u_i(t)$ is the value of the fortune in experiment number i, $i = 0, \ldots, N-1$, the mean evolution of the fortune is

$$\bar{u}(t) = \frac{1}{N}\sum_{i=0}^{N-1} u_i(t),$$

and the standard deviation is

$$s(t) = \sqrt{\frac{1}{N-1}\left(-(\bar{u}(t))^2 + \sum_{i=0}^{N-1}(u_i(t))^2\right)}.$$

Suppose $u_i(t)$ is stored in an array u. The mean and the standard deviation of the fortune is most efficiently computed by using two accumulation arrays, `sum_u` and `sum_u2`, and performing `sum_u += u` and `sum_u2 += u**2` after every experiment. This technique avoids storing all the $u_i(t)$ time series for computing the statistics.
Filename: `random_interest`.

Exercise 4.20: Simulate a population in a changing environment
We shall study a population modeled by (4.3) where the environment, represented by r and f, undergoes changes with time.

a) Assume that there is a sudden drop (increase) in the birth (death) rate at time $t = t_r$, because of limited nutrition or food supply:

$$r(t) = \begin{cases} \varrho, & t < t_r, \\ \varrho - A, & t \geq t_r. \end{cases}$$

This drop in population growth is compensated by a sudden net immigration at time $t_f > t_r$:

$$f(t) = \begin{cases} 0, & t < t_f, \\ f_0, & t \geq t_a. \end{cases}$$

Start with ϱ and make $A > \varrho$. Experiment with these and other parameters to illustrate the interplay of growth and decay in such a problem.

b) Now we assume that the environmental conditions changes periodically with time so that we may take

$$r(t) = \varrho + A \sin\left(\frac{2\pi}{P}t\right).$$

That is, the combined birth and death rate oscillates around ϱ with a maximum change of $\pm A$ repeating over a period of length P in time. Set $f = 0$ and experiment with the other parameters to illustrate typical features of the solution.

Filename: `population.py`.

Exercise 4.21: Simulate logistic growth

Solve the logistic ODE (4.4) using a Crank–Nicolson scheme where $(u^{n+\frac{1}{2}})^2$ is approximated by a *geometric mean*:

$$(u^{n+\frac{1}{2}})^2 \approx u^{n+1}u^n .$$

This trick makes the discrete equation linear in u^{n+1}.

Filename: `logistic_CN`.

Exercise 4.22: Rederive the equation for continuous compound interest

The ODE model (4.7) was derived under the assumption that r was constant. Perform an alternative derivation without this assumption: 1) start with (4.5); 2) introduce a time step Δt instead of m: $\Delta t = 1/m$ if t is measured in years; 3) divide by Δt and take the limit $\Delta t \to 0$. Simulate a case where the inflation is at a constant level I percent per year and the interest rate oscillates: $r = -I/2 + r_0 \sin(2\pi t)$. Compare solutions for $r_0 = I, 3I/2, 2I$.

Filename: `interest_modeling`.

Exercise 4.23: Simulate the deformation of a viscoelastic material

Stretching a rod made of polymer will cause deformations that are well described with a Kelvin–Voigt material model (4.49). At $t = 0$ we apply a constant force $\sigma = \sigma_0$, but at $t = t_1$, we remove the force so $\sigma = 0$. Compute numerically the corresponding strain (elongation divided by the rod's length) and visualize how it responds in time.

Hint To avoid finding proper values of the E and η parameters for a polymer, one can scale the problem. A common dimensionless time is $\bar{t} = tE/\eta$. Note that ε is already dimensionless by definition, but it takes on small values, say up to 0.1, so we introduce a scaling: $\bar{u} = 10\varepsilon$ such that $\bar{u}$ takes on values up to about unity.

Show that the material model then takes the form $\bar{u}' = -\bar{u} + 10\sigma(t)/E$. Work with the dimensionless force $F = 10\sigma(t)/E$, and let $F = 1$ for $\bar{t} \in (0, \bar{t}_1)$ and $F = 0$ for $\bar{t} \geq \bar{t}_1$. A possible choice of t_1 is the characteristic time η/E, which means that $\bar{t}_1 = 1$.

Filename: `KelvinVoigt`.

Scientific Software Engineering

5

Teaching material on scientific computing has traditionally been very focused on the mathematics and the applications, while details on how the computer is programmed to solve the problems have received little attention. Many end up writing as simple programs as possible, without being aware of much useful computer science technology that would increase the fun, efficiency, and reliability of the their scientific computing activities.

This chapter demonstrates a series of good practices and tools from modern computer science, using the simple mathematical problem $u' = -au$, $u(0) = I$, such that we minimize the mathematical details and can go more in depth with implementations. The goal is to increase the technological quality of computer programming and make it match the more well-established quality of the mathematics of scientific computing.

The conventions and techniques outlined here will save you a lot of time when you incrementally extend software over time from simpler to more complicated problems. In particular, you will benefit from many good habits:

- new code is added in a modular fashion to a library (modules),
- programs are run through convenient user interfaces,
- it takes one quick command to let all your code undergo heavy testing,
- tedious manual work with running programs is automated,
- your scientific investigations are reproducible,
- scientific reports with top quality typesetting are produced both for paper and electronic devices.

5.1 Implementations with Functions and Modules

All previous examples in this book have implemented numerical algorithms as Python functions. This is a good style that readers are expected to adopt. However, this author has experienced that many students and engineers are inclined to make "flat" programs, i.e., a sequence of statements without any use of functions, just to get the problem solved as quickly as possible. Since this programming style is so widespread, especially among people with MATLAB experience, we shall

H.P. Langtangen, *Finite Difference Computing with Exponential Decay Models*,
Lecture Notes in Computational Science and Engineering 110,
DOI 10.1007/978-3-319-29439-1_5

look at the weaknesses of flat programs and show how they can be *refactored* into more reusable programs based on functions.

5.1.1 Mathematical Problem and Solution Technique

We address the differential equation problem

$$u'(t) = -au(t), \quad t \in (0, T], \tag{5.1}$$

$$u(0) = I, \tag{5.2}$$

where a, I, and T are prescribed parameters, and $u(t)$ is the unknown function to be estimated. This mathematical model is relevant for physical phenomena featuring exponential decay in time, e.g., vertical pressure variation in the atmosphere, cooling of an object, and radioactive decay.

As we learned in Chap. 1.1.2, the time domain is discretized with points $0 = t_0 < t_1 \cdots < t_{N_t} = T$, here with a constant spacing Δt between the mesh points: $\Delta t = t_n - t_{n-1}$, $n = 1, \ldots, N_t$. Let u^n be the numerical approximation to the exact solution at t_n. A family of popular numerical methods are provided by the θ scheme,

$$u^{n+1} = \frac{1 - (1-\theta)a\Delta t}{1 + \theta a \Delta t} u^n, \tag{5.3}$$

for $n = 0, 1, \ldots, N_t - 1$. This formula produces the Forward Euler scheme when $\theta = 0$, the Backward Euler scheme when $\theta = 1$, and the Crank–Nicolson scheme when $\theta = 1/2$.

5.1.2 A First, Quick Implementation

Solving (5.3) in a program is very straightforward: just make a loop over n and evaluate the formula. The $u(t_n)$ values for discrete n can be stored in an array. This makes it easy to also plot the solution. It would be natural to also add the exact solution curve $u(t) = Ie^{-at}$ to the plot.

Many have programming habits that would lead them to write a simple program like this:

```
from numpy import *
from matplotlib.pyplot import *

A = 1
a = 2
T = 4
dt = 0.2
N = int(round(T/dt))
y = zeros(N+1)
t = linspace(0, T, N+1)
theta = 1
y[0] = A
for n in range(0, N):
    y[n+1] = (1 - (1-theta)*a*dt)/(1 + theta*dt*a)*y[n]
```

```
y_e = A*exp(-a*t) - y
error = y_e - y
E = sqrt(dt*sum(error**2))
print 'Norm of the error: %.3E' % E
plot(t, y, 'r--o')
t_e = linspace(0, T, 1001)
y_e = A*exp(-a*t_e)
plot(t_e, y_e, 'b-')
legend(['numerical, theta=%g' % theta, 'exact'])
xlabel('t')
ylabel('y')
show()
```

This program is easy to read, and as long as it is correct, many will claim that it has sufficient quality. Nevertheless, the program suffers from two serious flaws:

1. The notation in the program does not correspond *exactly* to the notation in the mathematical problem: the solution is called y and corresponds to u in the mathematical description, the variable `A` corresponds to the mathematical parameter I, `N` in the program is called N_t in the mathematics.
2. There are no comments in the program.

These kind of flaws quickly become crucial if present in code for complicated mathematical problems and code that is meant to be extended to other problems.

We also note that the program is *flat* in the sense that it does not contain functions. Usually, this is a bad habit, but let us first correct the two mentioned flaws.

5.1.3 A More Decent Program

A code of better quality arises from fixing the notation and adding comments:

```
from numpy import *
from matplotlib.pyplot import *

I = 1
a = 2
T = 4
dt = 0.2
Nt = int(round(T/dt))     # no of time intervals
u = zeros(Nt+1)           # array of u[n] values
t = linspace(0, T, Nt+1)  # time mesh
theta = 1                 # Backward Euler method

u[0] = I                  # assign initial condition
for n in range(0, Nt):    # n=0,1,...,Nt-1
    u[n+1] = (1 - (1-theta)*a*dt)/(1 + theta*dt*a)*u[n]

# Compute norm of the error
u_e = I*exp(-a*t) - u     # exact u at the mesh points
error = u_e - u
E = sqrt(dt*sum(error**2))
print 'Norm of the error: %.3E' % E

# Compare numerical (u) and exact solution (u_e) in a plot
plot(t, u, 'r--o')
t_e = linspace(0, T, 1001)       # very fine mesh for u_e
u_e = I*exp(-a*t_e)
```

```
plot(t_e, u_e, 'b-')
legend(['numerical, theta=%g' % theta, 'exact'])
xlabel('t')
ylabel('u')
show()
```

Comments in a program There is obviously not just one way to comment a program, and opinions may differ as to what code should be commented. The guiding principle is, however, that comments should make the program easy to understand for the human eye. Do not comment obvious constructions, but focus on ideas and ("what happens in the next statements?") and on explaining code that can be found complicated.

Refactoring into functions At first sight, our updated program seems like a good starting point for playing around with the mathematical problem: we can just change parameters and rerun. Although such edit-and-rerun sessions are good for initial exploration, one will soon extend the experiments and start developing the code further. Say we want to compare $\theta = 0, 1, 0.5$ in the same plot. This extension requires changes all over the code and quickly leads to errors. To do something serious with this program, we have to break it into smaller pieces and make sure each piece is well tested, and ensure that the program is sufficiently general and can be reused in new contexts without changes. The next natural step is therefore to isolate the numerical computations and the visualization in separate Python functions. Such a rewrite of a code, without essentially changing the functionality, but just improve the quality of the code, is known as *refactoring*. After quickly putting together and testing a program, the next step is to refactor it so it becomes better prepared for extensions.

Program file vs IDE vs notebook There are basically three different ways of working with Python code:

1. One writes the code in a file, using a text editor (such as Emacs or Vim) and runs it in a terminal window.
2. One applies an *Integrated Development Environment* (the simplest is IDLE, which comes with standard Python) containing a graphical user interface with an editor and an element where Python code can be run.
3. One applies the Jupyter Notebook (previously known as IPython Notebook), which offers an interactive environment for Python code where plots are automatically inserted after the code, see Fig. 5.1.

It appears that method 1 and 2 are quite equivalent, but the notebook encourages more experimental code and therefore also flat programs. Consequently, notebook users will normally need to think more about refactoring code and increase the use of functions after initial experimentation.

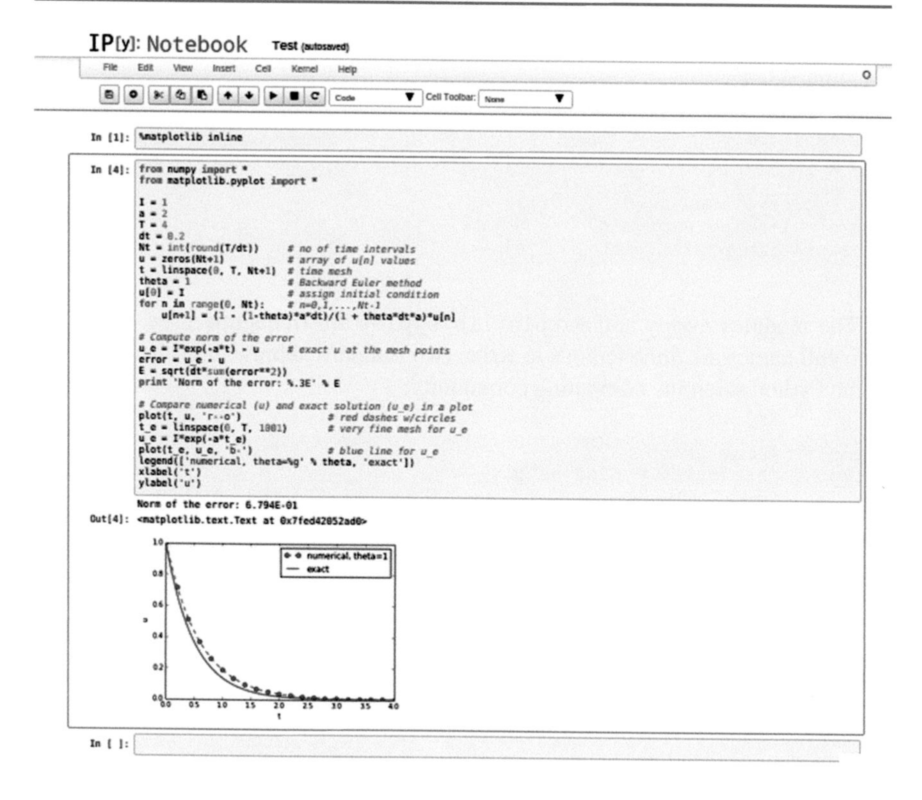

Fig. 5.1 Experimental code in a notebook

5.1.4 Prefixing Imported Functions by the Module Name

Import statements of the form `from module import *` import *all* functions and variables in `module.py` into the current file. This is often referred to as "import star", and many find this convenient, but it is not considered as a good programming style in Python. For example, when doing

```
from numpy import *
from matplotlib.pyplot import *
```

we get mathematical functions like `sin` and `exp` as well as MATLAB-style functions like `linspace` and `plot`, which can be called by these well-known names. Unfortunately, it sometimes becomes confusing to know where a particular function comes from, i.e., what modules you need to import. Is a desired function from `numpy` or `matplotlib.pyplot`? Or is it our own function? These questions are easy to answer if functions in modules are prefixed by the module name. Doing an additional `from math import *` is really crucial: now `sin`, `cos`, and other mathematical functions are imported and their names hide those previously imported from `numpy`. That is, `sin` is now a sine function that accepts a `float` argument, not an array.

Doing the import such that module functions must have a prefix is generally recommended:

```
import numpy
import matplotlib.pyplot

t = numpy.linspace(0, T, Nt+1)
u_e = I*numpy.exp(-a*t)
matplotlib.pyplot.plot(t, u_e)
```

The modules `numpy` and `matplotlib.pyplot` are frequently used, and since their full names are quite tedious to write, two standard abbreviations have evolved in the Python scientific computing community:

```
import numpy as np
import matplotlib.pyplot as plt

t = np.linspace(0, T, Nt+1)
u_e = I*np.exp(-a*t)
plt.plot(t, u_e)
```

The downside of prefixing functions by the module name is that mathematical expressions like $e^{-at}\sin(2\pi t)$ get cluttered with module names,

```
numpy.exp(-a*t)*numpy.sin(2(numpy.pi*t)
# or
np.exp(-a*t)*np.sin(2*np.pi*t)
```

Such an expression looks like `exp(-a*t)*sin(2*pi*t)` in most other programming languages. Similarly, `np.linspace` and `plt.plot` look less familiar to people who are used to MATLAB and who have not adopted Python's prefix style. Whether to do `from module import *` or `import module` depends on personal taste and the problem at hand. In these writings we use `from module import *` in more basic, shorter programs where similarity with MATLAB could be an advantage. However, in reusable modules we prefix calls to module functions by their function name, *or* do explicit import of the needed functions:

```
from numpy import exp, sum, sqrt

def u_exact(t, I, a):
    return I*exp(-a*t)

error = u_exact(t, I, a) - u
E = sqrt(dt*sum(error**2))
```

Prefixing module functions or not?

It can be advantageous to do a combination: mathematical functions in formulas are imported without prefix, while module functions in general are called with a prefix. For the `numpy` package we can do

```
import numpy as np
from numpy import exp, sum, sqrt
```

such that mathematical expression can apply `exp`, `sum`, and `sqrt` and hence look as close to the mathematical formulas as possible (without a disturbing prefix). Other calls to `numpy` function are done with the prefix, as in `np.linspace`.

5.1.5 Implementing the Numerical Algorithm in a Function

The solution formula (5.3) is completely general and should be available as a Python function `solver` with all input data as function arguments and all output data returned to the calling code. With this `solver` function we can solve all types of problems (5.1)–(5.2) by an easy-to-read one-line statement:

```
u, t = solver(I=1, a=2, T=4, dt=0.2, theta=0.5)
```

Refactoring the numerical method in the previous flat program in terms of a `solver` function and prefixing calls to module functions by the module name leads to this code:

```
def solver(I, a, T, dt, theta):
    """Solve u'=-a*u, u(0)=I, for t in (0,T] with steps of dt."""
    dt = float(dt)                  # avoid integer division
    Nt = int(round(T/dt))           # no of time intervals
    T = Nt*dt                       # adjust T to fit time step dt
    u = np.zeros(Nt+1)              # array of u[n] values
    t = np.linspace(0, T, Nt+1)     # time mesh

    u[0] = I                    # assign initial condition
    for n in range(0, Nt):      # n=0,1,...,Nt-1
        u[n+1] = (1 - (1-theta)*a*dt)/(1 + theta*dt*a)*u[n]
    return u, t
```

Tip: Always use a doc string to document a function!
Python has a convention for documenting the purpose and usage of a function in a *doc string*: simply place the documentation in a one- or multi-line triple-quoted string right after the function header.

Be careful with unintended integer division!
Note that we in the `solver` function explicitly covert `dt` to a `float` object. If not, the updating formula for `u[n+1]` may evaluate to zero because of integer division when `theta`, a, and `dt` are integers!

5.1.6 Do not Have Several Versions of a Code

One of the most serious flaws in computational work is to have several slightly different implementations of the same computational algorithms lying around in various program files. This is very likely to happen, because busy scientists often want to test a slight variation of a code to see what happens. A quick copy-and-edit does the task, but such quick hacks tend to survive. When a real correction is needed in the implementation, it is difficult to ensure that the correction is done in all relevant files. In fact, this is a general problem in programming, which has led to an important principle.

The DRY principle: Don't repeat yourself!

When implementing a particular functionality in a computer program, make sure this functionality and its variations are implemented in just one piece of code. That is, if you need to revise the implementation, there should be *one and only one* place to edit. It follows that you should never duplicate code (don't repeat yourself!), and code snippets that are similar should be factored into one piece (function) and parameterized (by function arguments).

The DRY principle means that our `solver` function should not be copied to a new file if we need some modifications. Instead, we should try to extend `solver` such that the new and old needs are met by a single function. Sometimes this process requires a new refactoring, but having a numerical method in one and only one place is a great advantage.

5.1.7 Making a Module

As soon as you start making Python functions in a program, you should make sure the program file fulfills the requirement of a module. This means that you can import and reuse your functions in other programs too. For example, if our `solver` function resides in a module file `decay.py`, another program may reuse of the function either by

```
from decay import solver
u, t = solver(I=1, a=2, T=4, dt=0.2, theta=0.5)
```

or by a slightly different import statement, combined with a subsequent prefix of the function name by the name of the module:

```
import decay
u, t = decay.solver(I=1, a=2, T=4, dt=0.2, theta=0.5)
```

The requirements for a program file to also qualify for a module are simple:

1. The filename without `.py` must be a valid Python variable name.
2. The main program must be executed (through statements or a function call) in the *test block*.

The *test block* is normally placed at the end of a module file:

```
if __name__ == '__main__':
    # Statements
```

When the module file is executed as a stand-alone program, the if test is true and the indented statements are run. If the module file is imported, however, `__name__` equals the module name and the test block is not executed.

To demonstrate the difference, consider the trivial module file `hello.py` with one function and a call to this function as main program:

```
def hello(arg='World!'):
    print 'Hello, ' + arg

if __name__ == '__main__':
    hello()
```

Without the test block, the code reads

```
def hello(arg='World!'):
    print 'Hello, ' + arg

hello()
```

With this latter version of the file, any attempt to import `hello` will, at the same time, execute the call `hello()` and hence write "Hello, World!" to the screen. Such output is not desired when importing a module! To make import and execution of code independent for another program that wants to use the function `hello`, the module `hello` must be written with a test block. Furthermore, running the file itself as `python hello.py` will make the block active and lead to the desired printing.

All coming functions are placed in a module!

The many functions to be explained in the following text are put in one module file `decay.py`[1].

What more than the `solver` function is needed in our `decay` module to do everything we did in the previous, flat program? We need import statements for `numpy` and `matplotlib` as well as another function for producing the plot. It can also be convenient to put the exact solution in a Python function. Our module `decay.py` then looks like this:

```
import numpy as np
import matplotlib.pyplot as plt

def solver(I, a, T, dt, theta):
    ...

def u_exact(t, I, a):
    return I*np.exp(-a*t)

def experiment_compare_numerical_and_exact():
    I = 1;  a = 2;  T = 4;  dt = 0.4;  theta = 1
    u, t = solver(I, a, T, dt, theta)

    t_e = np.linspace(0, T, 1001)       # very fine mesh for u_e
    u_e = u_exact(t_e, I, a)

    plt.plot(t,   u,   'r--o')       # dashed red line with circles
    plt.plot(t_e, u_e, 'b-')         # blue line for u_e
    plt.legend(['numerical, theta=%g' % theta, 'exact'])
    plt.xlabel('t')
    plt.ylabel('u')
    plotfile = 'tmp'
    plt.savefig(plotfile + '.png');  plt.savefig(plotfile + '.pdf')
```

[1] http://tinyurl.com/ofkw6kc/softeng/decay.py

```
    error = u_exact(t, I, a) - u
    E = np.sqrt(dt*np.sum(error**2))
    print 'Error norm:', E

if __name__ == '__main__':
    experiment_compare_numerical_and_exact()
```

We could consider doing `from numpy import exp, sqrt, sum` to make the mathematical expressions with these functions closer to the mathematical formulas, but here we employed the prefix since the formulas are so simple and easy to read.

This module file does exactly the same as the previous, flat program, but now it becomes much easier to extend the code with more functions that produce other plots, other experiments, etc. Even more important, though, is that the numerical algorithm is coded and tested once and for all in the `solver` function, and any need to solve the mathematical problem is a matter of one function call.

5.1.8 Example on Extending the Module Code

Let us specifically demonstrate one extension of the flat program in Sect. 5.1.2 that would require substantial editing of the flat code (Sect. 5.1.3), while in a structured module (Sect. 5.1.7), we can simply add a new function without affecting the existing code.

Our example that illustrates the extension is to make a comparison between the numerical solutions for various schemes (θ values) and the exact solution:

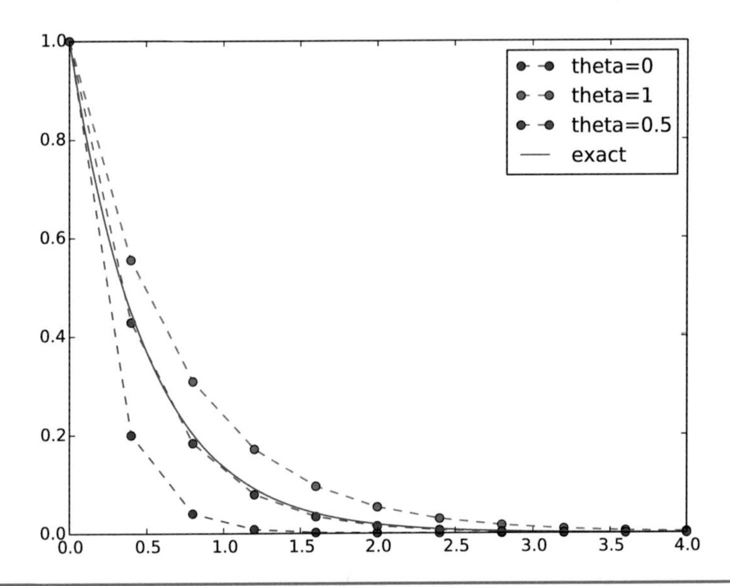

Wait a minute!

Look at the flat program in Sect. 5.1.2, and try to imagine which edits that are required to solve this new problem.

With the `solver` function at hand, we can simply create a function with a loop over `theta` values and add the necessary plot statements:

```
def experiment_compare_schemes():
    """Compare theta=0,1,0.5 in the same plot."""
    I = 1;  a = 2;  T = 4;  dt = 0.4
    legends = []
    for theta in [0, 1, 0.5]:
        u, t = solver(I, a, T, dt, theta)
        plt.plot(t, u, '--o')
        legends.append('theta=%g' % theta)
    t_e = np.linspace(0, T, 1001)        # very fine mesh for u_e
    u_e = u_exact(t_e, I, a)
    plt.plot(t_e, u_e, 'b-')
    legends.append('exact')
    plt.legend(legends, loc='upper right')
    plotfile = 'tmp'
    plt.savefig(plotfile + '.png');  plt.savefig(plotfile + '.pdf')
```

A call to this experiment_compare_schemes function must be placed in the test block, or you can run the program from IPython instead:

```
In[1]: from decay import *

In[2]: experiment_compare_schemes()
```

We do not present how the flat program from Sect. 5.1.3 must be refactored to produce the desired plots, but simply state that the danger of introducing bugs is significantly larger than when just writing an additional function in the decay module.

5.1.9 Documenting Functions and Modules

We have already emphasized the importance of documenting functions with a doc string (see Sect. 5.1.5). Now it is time to show how doc strings should be structured in order to take advantage of the documentation utilities in the numpy module. The idea is to follow a convention that in itself makes a good pure text doc string in the terminal window and at the same time can be translated to beautiful HTML manuals for the web.

The conventions for numpy style doc strings are well documented[2], so here we just present a basic example that the reader can adopt. Input arguments to a function are listed under the heading Parameters, while returned values are listed under Returns. It is a good idea to also add an Examples section on the usage of the function. More complicated software may have additional sections, see pydoc numpy.load for an example. The markup language available for doc strings is Sphinx-extended reStructuredText. The example below shows typical constructs: 1) how inline mathematics is written with the :math: directive, 2) how arguments to the functions are referred to using single backticks (inline monospace font for code applies double backticks), and 3) how arguments and return values are listed with types and explanation.

[2] https://github.com/numpy/numpy/blob/master/doc/HOWTO_DOCUMENT.rst.txt

```
def solver(I, a, T, dt, theta):
    """
    Solve :math:`u'=-au` with :math:`u(0)=I` for :math:`t \in (0,T]`
    with steps of `dt` and the method implied by `theta`.

    Parameters
    ----------
    I: float
        Initial condition.
    a: float
        Parameter in the differential equation.
    T: float
        Total simulation time.
    theta: float, int
        Parameter in the numerical scheme. 0 gives
        Forward Euler, 1 Backward Euler, and 0.5
        the centered Crank-Nicolson scheme.

    Returns
    -------
    `u`: array
        Solution array.
    `t`: array
        Array with time points corresponding to `u`.

    Examples
    --------
    Solve :math:`u' = -\frac{1}{2}u, u(0)=1.5`
    with the Crank-Nicolson method:

    >>> u, t = solver(I=1.5, a=0.5, T=9, theta=0.5)
    >>> import matplotlib.pyplot as plt
    >>> plt.plot(t, u)
    >>> plt.show()
    """
```

If you follow such doc string conventions in your software, you can easily produce nice manuals that meet the standard expected within the Python scientific computing community.

Sphinx[3] requires quite a number of manual steps to prepare a manual, so it is recommended to use a premade script[4] to automate the steps. (By default, the script generates documentation for all `*.py` files in the current directory. You need to do a `pip install` of `sphinx` and `numpydoc` to make the script work.) Figure 5.2 provides an example of what the above doc strings look like when Sphinx has transformed them to HTML.

5.1.10 Logging Intermediate Results

Sometimes one may wish that a simulation program could write out intermediate results for inspection. This could be accomplished by a (global) `verbose` variable and code like

```
if verbose >= 2:
    print 'u[%d]=%g' % (i, u[i])
```

[3] http://sphinx-doc.org/
[4] http://tinyurl.com/ofkw6kc/softeng/make_sphinx_api.py

	T: float
	Total simulation time.
	theta: float, int
	Parameter in the numerical scheme. 0 gives Forward Euler, 1 Backward Euler, and 0.5 the centered Crank-Nicolson scheme.
Returns:	u: array
	Solution array.
	t: array
	Array with time points corresponding to *u*.

Examples

Solve $u' = -\frac{1}{2}u, u(0) = 1.5$ with the Crank-Nicolson method:

```
>>> u, t = solver(I=1.5, a=0.5, T=9, theta=0.5)
>>> import matplotlib.pyplot as plt
>>> plt.plot(t, u)
>>> plt.show()
```

Fig. 5.2 Example on Sphinx API manual in HTML

The professional way to do report intermediate results and problems is, however, to use a *logger*. This is an object that writes messages to a log file. The messages are classified as debug, info, and warning.

Introductory example Here is a simple example using defining a logger, using Python's `logging` module:

```
import logging
# Configure logger
logging.basicConfig(
    filename='myprog.log', filemode='w', level=logging.WARNING,
    format='%(asctime)s - %(levelname)s - %(message)s',
    datefmt='%m/%d/%Y %I:%M:%S %p')
# Perform logging
logging.info('Here is some general info.')
logging.warning('Here is a warning.')
logging.debug('Here is some debugging info.')
logging.critical('Dividing by zero!')
logging.error('Encountered an error.')
```

Running this program gives the following output in the log file `myprog.log`:

```
09/26/2015 09:25:10 AM - INFO - Here is some general info.
09/26/2015 09:25:10 AM - WARNING - Here is a warning.
09/26/2015 09:25:10 AM - CRITICAL - Dividing by zero!
09/26/2015 09:25:10 AM - ERROR - Encountered an error.
```

The logger has different *levels* of messages, ordered as *critical*, *error*, *warning*, *info*, and *debug*. The `level` argument to `logging.basicConfig` sets the level and thereby determines what the logger will print to the file: all messages at the specified *and lower* levels are printed. For example, in the above example we set the level to be *info*, and therefore the critical, error, warning, and info messages were printed, but not the debug message. Setting level to debug (`logging.DEBUG`) prints all messages, while level *critical* prints only the critical messages.

The `filemode` argument is set to `w` such that any existing log file is overwritten (the default is `a`, which means append new messages to an existing log file, but this is seldom what you want in mathematical computations).

The messages are preceded by the date and time and the level of the message. This output is governed by the `format` argument: `asctime` is the date and time, `levelname` is the name of the message level, and `message` is the message itself. Setting `format='%(message)s'` ensures that just the message and nothing more is printed on each line. The `datefmt` string specifies the formatting of the date and time, using the rules of the `time.strftime`[5] function.

Using a logger in our solver function Let us let a logger write out intermediate results and some debugging results in the `solver` function. Such messages are useful for monitoring the simulation and for debugging it, respectively.

```
def solver_with_logging(I, a, T, dt, theta):
    """Solve u'=-a*u, u(0)=I, for t in (0,T] with steps of dt."""
    dt = float(dt)               # avoid integer division
    Nt = int(round(T/dt))        # no of time intervals
    T = Nt*dt                    # adjust T to fit time step dt
    u = np.zeros(Nt+1)           # array of u[n] values
    t = np.linspace(0, T, Nt+1)  # time mesh
    logging.debug('solver: dt=%g, Nt=%g, T=%g' % (dt, Nt, T))

    u[0] = I                  # assign initial condition
    for n in range(0, Nt):    # n=0,1,...,Nt-1
        u[n+1] = (1 - (1-theta)*a*dt)/(1 + theta*dt*a)*u[n]

        logging.info('u[%d]=%g' % (n, u[n]))
        logging.debug('1 - (1-theta)*a*dt: %g, %s' %
                      (1-(1-theta)*a*dt,
                       str(type(1-(1-theta)*a*dt))[7:-2]))
        logging.debug('1 + theta*dt*a: %g, %s' %
                      (1 + theta*dt*a,
                       str(type(1 + theta*dt*a))[7:-2]))
    return u, t
```

The application code that calls `solver_with_logging` needs to configure the logger. The `decay` module offers a default configure function:

```
import logging

def configure_basic_logger():
    logging.basicConfig(
        filename='decay.log', filemode='w', level=logging.DEBUG,
        format='%(asctime)s - %(levelname)s - %(message)s',
        datefmt='%Y.%m.%d %I:%M:%S %p')
```

[5] https://docs.python.org/2/library/time.html#time.strftime

If the user of a library does not configure a logger or call this configure function, the library should prevent error messages by declaring a default logger that does nothing:

```
import logging
logging.getLogger('decay').addHandler(logging.NullHandler())
```

We can run the new solver function with logging in a shell:

```
>>> import decay
>>> decay.configure_basic_logger()
>>> u, t = decay.solver_with_logging(I=1, a=0.5, T=10, \
          dt=0.5, theta=0.5)
```

During this execution, each logging message is appended to the log file. Suppose we add some pause (`time.sleep(2)`) at each time level such that the execution takes some time. In another terminal window we can then monitor the evolution of `decay.log` and the simulation by the `tail -f` Unix command:

```
Terminal> tail -f decay.log
2015.09.26 05:37:41 AM - INFO - u[0]=1
2015.09.26 05:37:41 AM - INFO - u[1]=0.777778
2015.09.26 05:37:41 AM - INFO - u[2]=0.604938
2015.09.26 05:37:41 AM - INFO - u[3]=0.470508
2015.09.26 05:37:41 AM - INFO - u[4]=0.36595
2015.09.26 05:37:41 AM - INFO - u[5]=0.284628
```

Especially in simulation where each time step demands considerable CPU time (minutes, hours), it can be handy to monitor such a log file to see the evolution of the simulation.

If we want to look more closely into the numerator and denominator of the formula for u^{n+1}, we can change the logging level to `level=logging.DEBUG` and get output in `decay.log` like

```
2015.09.26 05:40:01 AM - DEBUG - solver: dt=0.5, Nt=20, T=10
2015.09.26 05:40:01 AM - INFO - u[0]=1
2015.09.26 05:40:01 AM - DEBUG - 1 - (1-theta)*a*dt: 0.875, float
2015.09.26 05:40:01 AM - DEBUG - 1 + theta*dt*a: 1.125, float
2015.09.26 05:40:01 AM - INFO - u[1]=0.777778
2015.09.26 05:40:01 AM - DEBUG - 1 - (1-theta)*a*dt: 0.875, float
2015.09.26 05:40:01 AM - DEBUG - 1 + theta*dt*a: 1.125, float
2015.09.26 05:40:01 AM - INFO - u[2]=0.604938
2015.09.26 05:40:01 AM - DEBUG - 1 - (1-theta)*a*dt: 0.875, float
2015.09.26 05:40:01 AM - DEBUG - 1 + theta*dt*a: 1.125, float
2015.09.26 05:40:01 AM - INFO - u[3]=0.470508
2015.09.26 05:40:01 AM - DEBUG - 1 - (1-theta)*a*dt: 0.875, float
2015.09.26 05:40:01 AM - DEBUG - 1 + theta*dt*a: 1.125, float
2015.09.26 05:40:01 AM - INFO - u[4]=0.36595
2015.09.26 05:40:01 AM - DEBUG - 1 - (1-theta)*a*dt: 0.875, float
2015.09.26 05:40:01 AM - DEBUG - 1 + theta*dt*a: 1.125, float
```

Logging can be much more sophisticated than shown above. One can, e.g., output critical messages to the screen and all messages to a file. This requires more code as demonstrated in the Logging Cookbook[6].

[6] https://docs.python.org/2/howto/logging-cookbook.html

5.2 User Interfaces

It is good programming practice to let programs read input from some *user interface*, rather than requiring users to *edit* parameter values in the source code. With effective user interfaces it becomes easier and safer to apply the code for scientific investigations and in particular to automate large-scale investigations by other programs (see Sect. 5.6).

Reading input data can be done in many ways. We have to decide on the functionality of the user interface, i.e., how we want to operate the program when providing input. Thereafter, we use appropriate tools to implement the particular user interface. There are four basic types of user interface, listed here according to implementational complexity, from lowest to highest:

1. Questions and answers in the terminal window
2. Command-line arguments
3. Reading data from files
4. Graphical user interfaces (GUIs)

Personal preferences of user interfaces differ substantially, and it is difficult to present recommendations or pros and cons. Alternatives 2 and 4 are most popular and will be addressed next. The goal is to make it easy for the user to set physical and numerical parameters in our `decay.py` program. However, we use a little toy program, called `prog.py`, as introductory example:

```
delta = 0.5
p = 2
from math import exp
result = delta*exp(-p)
print result
```

The essential content is that `prog.py` has two input parameters: `delta` and `p`. A user interface will replace the first two assignments to `delta` and `p`.

5.2.1 Command-Line Arguments

The command-line arguments are all the words that appear on the command line after the program name. Running a program `prog.py` as `python prog.py arg1 arg2` means that there are two command-line arguments (separated by white space): `arg1` and `arg2`. Python stores all command-line arguments in a special list `sys.argv`. (The name `argv` stems from the C language and stands for "argument values". In C there is also an integer variable called `argc` reflecting the number of arguments, or "argument counter". A lot of programming languages have adopted the variable name `argv` for the command-line arguments.) Here is an example on a program `what_is_sys_argv.py` that can show us what the command-line arguments are

```
import sys
print sys.argv
```

A sample run goes like

```
Terminal> python what_is_sys_argv.py 5.0 'two words' -1E+4
['what_is_sys_argv.py', '5.0', 'two words', '-1E+4']
```

We make two observations:

- `sys.argv[0]` is the name of the program, and the sublist `sys.argv[1:]` contains all the command-line arguments.
- Each command-line argument is available as a string. A conversion to `float` is necessary if we want to compute with the numbers 5.0 and 10^4.

There are, in principle, two ways of programming with command-line arguments in Python:

- **Positional arguments:** Decide upon a sequence of parameters on the command line and read their values directly from the `sys.argv[1:]` list.
- **Option-value pairs:** Use `-option value` on the command line to replace the default value of an input parameter `option` by `value` (and utilize the `argparse.ArgumentParser` tool for implementation).

Suppose we want to run some program `prog.py` with specification of two parameters `p` and `delta` on the command line. With positional command-line arguments we write

```
Terminal> python prog.py 2 0.5
```

and must know that the first argument 2 represents `p` and the next `0.5` is the value of `delta`. With option-value pairs we can run

```
Terminal> python prog.py --delta 0.5 --p 2
```

Now, both `p` and `delta` are supposed to have default values in the program, so we need to specify only the parameter that is to be changed from its default value, e.g.,

```
Terminal> python prog.py --p 2          # p=2, default delta
Terminal> python prog.py --delta 0.7    # delta-0.7, default a
Terminal> python prog.py                # default a and delta
```

How do we extend the `prog.py` code for positional arguments and option-value pairs? Positional arguments require very simple code:

```
import sys
p = float(sys.argv[1])
delta = float(sys.argv[2])

from math import exp
result = delta*exp(-p)
print result
```

If the user forgets to supply two command-line arguments, Python will raise an `IndexError` exception and produce a long error message. To avoid that, we should use a `try-except` construction:

```
import sys
try:
    p = float(sys.argv[1])
    delta = float(sys.argv[2])
except IndexError:
    print 'Usage: %s p delta' % sys.argv[0]
    sys.exit(1)

from math import exp
result = delta*exp(-p)
print result
```

Using `sys.exit(1)` aborts the program. The value 1 (actually any value different from 0) notifies the operating system that the program failed.

Command-line arguments are strings!

Note that all elements in `sys.argv` are string objects. If the values are used in mathematical computations, we need to explicitly convert the strings to numbers.

Option-value pairs requires more programming and is actually better explained in a more comprehensive example below. Minimal code for our `prog.py` program reads

```
import argparse
parser = argparse.ArgumentParser()
parser.add_argument('--p', default=1.0)
parser.add_argument('--delta', default=0.1)

args = parser.parse_args()
p = args.p
delta = args.delta

from math import exp
result = delta*exp(-p)
print result
```

Because the default values of `delta` and `p` are float numbers, the `args.delta` and `args.p` variables are automatically of type `float`.

Our next task is to use these basic code constructs to equip our `decay.py` module with command-line interfaces.

5.2.2 Positional Command-Line Arguments

For our `decay.py` module file, we want to include functionality such that we can read I, a, T, θ, and a range of Δt values from the command line. A plot is then to be made, comparing the different numerical solutions for different Δt values against the exact solution. The technical details of getting the command-line information into the program is covered in the next two sections.

The simplest way of reading the input parameters is to decide on their sequence on the command line and just index the `sys.argv` list accordingly. Say the sequence of input data for some functionality in `decay.py` is I, a, T, θ followed by an arbitrary number of Δt values. This code extracts these *positional* command-line arguments:

```
def read_command_line_positional():
    if len(sys.argv) < 6:
        print 'Usage: %s I a T on/off BE/FE/CN dt1 dt2 dt3 ...' % \
              sys.argv[0]; sys.exit(1)  # abort

    I = float(sys.argv[1])
    a = float(sys.argv[2])
    T = float(sys.argv[3])
    theta = float(sys.argv[4])
    dt_values = [float(arg) for arg in sys.argv[5:]]

    return I, a, T, theta, dt_values
```

Note that we may use a `try-except` construction instead of the if test.

A run like

```
Terminal
Terminal> python decay.py 1 0.5 4 0.5 1.5 0.75 0.1
```

results in

```
sys.argv = ['decay.py', '1', '0.5', '4', '0.5', '1.5', '0.75', '0.1']
```

and consequently the assignments `I=1.0`, `a=0.5`, `T=4.0`, `thet=0.5`, and `dt_values = [1.5, 0.75, 0.1]`.

Instead of specifying the θ value, we could be a bit more sophisticated and let the user write the name of the scheme: `BE` for Backward Euler, `FE` for Forward Euler, and `CN` for Crank–Nicolson. Then we must map this string to the proper θ value, an operation elegantly done by a dictionary:

```
scheme = sys.argv[4]
scheme2theta = {'BE': 1, 'CN': 0.5, 'FE': 0}
if scheme in scheme2theta:
    theta = scheme2theta[scheme]
else:
    print 'Invalid scheme name:', scheme; sys.exit(1)
```

Now we can do

Terminal

```
Terminal> python decay.py 1 0.5 4 CN 1.5 0.75 0.1
```

and get 'theta=0.5'in the code.

5.2.3 Option-Value Pairs on the Command Line

Now we want to specify option-value pairs on the command line, using `-I` for `I` (I), `-a` for `a` (a), `-T` for `T` (T), `-scheme` for the scheme name (`BE`, `FE`, `CN`), and `-dt` for the sequence of `dt` (Δt) values. Each parameter must have a sensible default value so that we specify the option on the command line only when the default value is not suitable. Here is a typical run:

Terminal

```
Terminal> python decay.py --I 2.5 --dt 0.1 0.2 0.01 --a 0.4
```

Observe the major advantage over positional command-line arguments: the input is much easier to read and much easier to write. With positional arguments it is easy to mess up the sequence of the input parameters and quite challenging to detect errors too, unless there are just a couple of arguments.

Python's `ArgumentParser` tool in the `argparse` module makes it easy to create a professional command-line interface to any program. The documentation of `ArgumentParser`[7] demonstrates its versatile applications, so we shall here just list an example containing the most basic features. It always pays off to use `ArgumentParser` rather than trying to manually inspect and interpret option-value pairs in `sys.argv`!

The use of `ArgumentParser` typically involves three steps:

```
import argparse
parser = argparse.ArgumentParser()

# Step 1: add arguments
parser.add_argument('--option_name', ...)

# Step 2: interpret the command line
args = parser.parse_args()

# Step 3: extract values
value = args.option_name
```

[7] http://docs.python.org/library/argparse.html

A function for setting up all the options is handy:

```
def define_command_line_options():
    import argparse
    parser = argparse.ArgumentParser()
    parser.add_argument(
        '--I', '--initial_condition', type=float,
        default=1.0, help='initial condition, u(0)',
        metavar='I')
    parser.add_argument(
        '--a', type=float, default=1.0,
        help='coefficient in ODE', metavar='a')
    parser.add_argument(
        '--T', '--stop_time', type=float,
        default=1.0, help='end time of simulation',
        metavar='T')
    parser.add_argument(
        '--scheme', type=str, default='CN',
        help='FE, BE, or CN')
    parser.add_argument(
        '--dt', '--time_step_values', type=float,
        default=[1.0], help='time step values',
        metavar='dt', nargs='+', dest='dt_values')
    return parser
```

Each command-line option is defined through the `parser.add_argument` method[8]. Alternative options, like the short `-I` and the more explaining version `--initial_condition` can be defined. Other arguments are `type` for the Python object type, a default value, and a help string, which gets printed if the command-line argument `-h` or `-help` is included. The `metavar` argument specifies the value associated with the option when the help string is printed. For example, the option for I has this help output:

Terminal

```
Terminal> python decay.py -h
  ...
  --I I, --initial_condition I
                        initial condition, u(0)
  ...
```

The structure of this output is

```
  --I metavar, --initial_condition metavar
                        help-string
```

Finally, the `-dt` option demonstrates how to allow for more than one value (separated by blanks) through the `nargs='+'` keyword argument. After the command line is parsed, we get an object where the values of the options are stored as attributes. The attribute name is specified by the `dist` keyword argument, which for the `-dt` option is `dt_values`. Without the `dest` argument, the value of an option `-opt` is stored as the attribute `opt`.

[8] We use the expression *method* here, because `parser` is a class variable and functions in classes are known as methods in Python and many other languages. Readers not familiar with class programming can just substitute this use of *method* by *function*.

The code below demonstrates how to read the command line and extract the values for each option:

```
def read_command_line_argparse():
    parser = define_command_line_options()
    args = parser.parse_args()
    scheme2theta = {'BE': 1, 'CN': 0.5, 'FE': 0}
    data = (args.I, args.a, args.T, scheme2theta[args.scheme],
            args.dt_values)
    return data
```

As seen, the values of the command-line options are available as attributes in `args`: `args.opt` holds the value of option `-opt`, unless we used the `dest` argument (as for `--dt_values`) for specifying the attribute name. The `args.opt` attribute has the object type specified by `type` (`str` by default).

The making of the plot is not dependent on whether we read data from the command line as positional arguments or option-value pairs:

```
def experiment_compare_dt(option_value_pairs=False):
    I, a, T, theta, dt_values = \
       read_command_line_argparse() if option_value_pairs else \
       read_command_line_positional()

    legends = []
    for dt in dt_values:
        u, t = solver(I, a, T, dt, theta)
        plt.plot(t, u)
        legends.append('dt=%g' % dt)
    t_e = np.linspace(0, T, 1001)       # very fine mesh for u_e
    u_e = u_exact(t_e, I, a)
    plt.plot(t_e, u_e, '--')
    legends.append('exact')
    plt.legend(legends, loc='upper right')
    plt.title('theta=%g' % theta)
    plotfile = 'tmp'
    plt.savefig(plotfile + '.png');  plt.savefig(plotfile + '.pdf')
```

5.2.4 Creating a Graphical Web User Interface

The Python package Parampool[9] can be used to automatically generate a web-based *graphical user interface* (GUI) for our simulation program. Although the programming technique dramatically simplifies the efforts to create a GUI, the forthcoming material on equipping our `decay` module with a GUI is quite technical and of significantly less importance than knowing how to make a command-line interface.

Making a compute function The first step is to identify a function that performs the computations and that takes the necessary input variables as arguments. This is called the *compute function* in Parampool terminology. The purpose of this function is to take values of I, a, T together with a sequence of Δt values and a sequence of θ and plot the numerical against the exact solution for each pair of $(\theta, \Delta t)$. The plots can be arranged as a table with the columns being scheme type (θ value) and

[9] https://github.com/hplgit/parampool

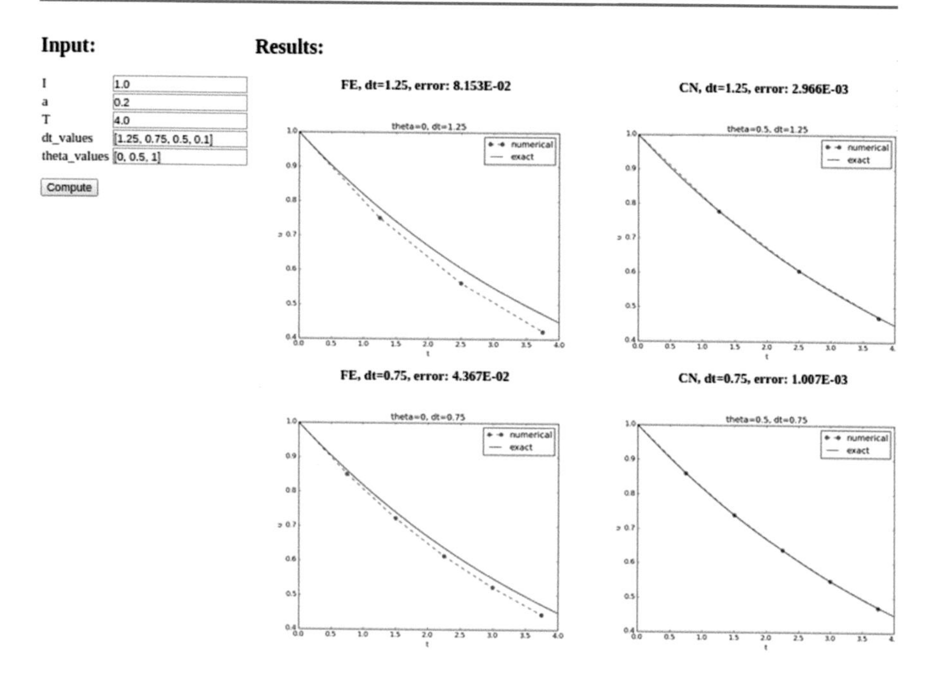

Fig. 5.3 Automatically generated graphical web interface

the rows reflecting the discretization parameter (Δt value). Figure 5.3 displays what the graphical web interface may look like after results are computed (there are 3×3 plots in the GUI, but only 2×2 are visible in the figure).

To tell Parampool what type of input data we have, we assign default values of the right type to all arguments in the compute function, here called `main_GUI`:

```
def main_GUI(I=1.0, a=.2, T=4.0,
             dt_values=[1.25, 0.75, 0.5, 0.1],
             theta_values=[0, 0.5, 1]):
```

The compute function must return the HTML code we want for displaying the result in a web page. Here we want to show a table of plots. Assume for now that the HTML code for one plot and the value of the norm of the error can be computed by some other function `compute4web`. The `main_GUI` function can then loop over Δt and θ values and put each plot in an HTML table. Appropriate code goes like

```
def main_GUI(I=1.0, a=.2, T=4.0,
             dt_values=[1.25, 0.75, 0.5, 0.1],
             theta_values=[0, 0.5, 1]):
    # Build HTML code for web page. Arrange plots in columns
    # corresponding to the theta values, with dt down the rows
    theta2name = {0: 'FE', 1: 'BE', 0.5: 'CN'}
    html_text = '<table>\n'
    for dt in dt_values:
        html_text += '<tr>\n'
        for theta in theta_values:
            E, html = compute4web(I, a, T, dt, theta)
            html_text += """
```

```
<td>
<center><b>%s, dt=%g, error: %.3E</b></center><br>
%s
</td>
""" % (theta2name[theta], dt, E, html)
        html_text += '</tr>\n'
    html_text += '</table>\n'
    return html_text
```

Making one plot is done in `compute4web`. The statements should be straightforward from earlier examples, but there is one new feature: we use a tool in Parampool to embed the PNG code for a plot file directly in an HTML image tag. The details are hidden from the programmer, who can just rely on relevant HTML code in the string `html_text`. The function looks like

```
def compute4web(I, a, T, dt, theta=0.5):
    """
    Run a case with the solver, compute error measure,
    and plot the numerical and exact solutions in a PNG
    plot whose data are embedded in an HTML image tag.
    """
    u, t = solver(I, a, T, dt, theta)
    u_e = u_exact(t, I, a)
    e = u_e - u
    E = np.sqrt(dt*np.sum(e**2))

    plt.figure()
    t_e = np.linspace(0, T, 1001)    # fine mesh for u_e
    u_e = u_exact(t_e, I, a)
    plt.plot(t,   u,   'r--o')
    plt.plot(t_e, u_e, 'b-')
    plt.legend(['numerical', 'exact'])
    plt.xlabel('t')
    plt.ylabel('u')
    plt.title('theta=%g, dt=%g' % (theta, dt))
    # Save plot to HTML img tag with PNG code as embedded data
    from parampool.utils import save_png_to_str
    html_text = save_png_to_str(plt, plotwidth=400)

    return E, html_text
```

Generating the user interface The web GUI is automatically generated by the following code, placed in the file `decay_GUI_generate.py`[10].

```
from parampool.generator.flask import generate
from decay import main_GUI
generate(main_GUI,
         filename_controller='decay_GUI_controller.py',
         filename_template='decay_GUI_view.py',
         filename_model='decay_GUI_model.py')
```

Running the `decay_GUI_generate.py` program results in three new files whose names are specified in the call to `generate`:

1. `decay_GUI_model.py` defines HTML widgets to be used to set input data in the web interface,

[10] http://tinyurl.com/ofkw6kc/softeng/decay_GUI_generate.py

2. `templates/decay_GUI_views.py` defines the layout of the web page,
3. `decay_GUI_controller.py` runs the web application.

We only need to run the last program, and there is no need to look into these files.

Running the web application The web GUI is started by

```
Terminal> python decay_GUI_controller.py
```

Open a web browser at the location `127.0.0.1:5000`. Input fields for `I`, `a`, `T`, `dt_values`, and `theta_values` are presented. Figure 5.3 shows a part of the resulting page if we run with the default values for the input parameters. With the techniques demonstrated here, one can easily create a tailored web GUI for a particular type of application and use it to interactively explore physical and numerical effects.

5.3 Tests for Verifying Implementations

Any module with functions should have a set of tests that can check the correctness of the implementations. There exists well-established procedures and corresponding tools for automating the execution of such tests. These tools allow large test sets to be run with a one-line command, making it easy to check that the software still works (as far as the tests can tell!). Here we shall illustrate two important software testing techniques: *doctest* and *unit testing*. The first one is Python specific, while unit testing is the dominating test technique in the software industry today.

5.3.1 Doctests

A doc string, the first string after the function header, is used to document the purpose of functions and their arguments (see Sect. 5.1.5). Very often it is instructive to include an example in the doc string on how to use the function. Interactive examples in the Python shell are most illustrative as we can see the output resulting from the statements and expressions. For example, in the `solver` function, we can include an example on calling this function and printing the computed `u` and `t` arrays:

```
def solver(I, a, T, dt, theta):
    """
    Solve u'=-a*u, u(0)=I, for t in (0,T] with steps of dt.

    >>> u, t = solver(I=0.8, a=1.2, T=1.5, dt=0.5, theta=0.5)
    >>> for n in range(len(t)):
    ...     print 't=%.1f, u=%.14f' % (t[n], u[n])
    t=0.0, u=0.80000000000000
    t=0.5, u=0.43076923076923
```

```
    t=1.0, u=0.23195266272189
    t=1.5, u=0.12489758761948
    """
    ...
```

When such interactive demonstrations are inserted in doc strings, Python's `doctest`[11] module can be used to automate running all commands in interactive sessions and compare new output with the output appearing in the doc string. All we have to do in the current example is to run the module file `decay.py` with

```
Terminal> python -m doctest decay.py
```

This command imports the `doctest` module, which runs all doctests found in the file and reports discrepancies between expected and computed output. Alternatively, the test block in a module may run all doctests by

```
if __name__ == '__main__':
    import doctest
    doctest.testmod()
```

Doctests can also be embedded in nose/pytest unit tests as explained in the next section.

Doctests prevent command-line arguments!

No additional command-line argument is allowed when running doctests. If your program relies on command-line input, make sure the doctests can be run *without* such input on the command line.

However, you can simulate command-line input by filling `sys.argv` with values, e.g.,

```
import sys; sys.argv = '--I 1.0 --a 5'.split()
```

The execution command above will report any problem if a test fails. As an illustration, let us alter the u value at `t=1.5` in the output of the doctest by replacing the last digit 8 by 7. This edit triggers a report:

Terminal

```
Terminal> python -m doctest decay.py
********************************************************
File "decay.py", line ...
Failed example:
    for n in range(len(t)):
        print 't=%.1f, u=%.14f' % (t[n], u[n])
Expected:
    t=0.0, u=0.80000000000000
    t=0.5, u=0.43076923076923
    t=1.0, u=0.23195266272189
    t=1.5, u=0.12489758761948
Got:
    t=0.0, u=0.80000000000000
    t=0.5, u=0.43076923076923
```

[11] http://docs.python.org/library/doctest.html

```
t=1.0, u=0.23195266272189
t=1.5, u=0.12489758761947
```

Pay attention to the number of digits in doctest results!
Note that in the output of `t` and `u` we write `u` with 14 digits. Writing all 16 digits is not a good idea: if the tests are run on different hardware, round-off errors might be different, and the `doctest` module detects that the numbers are not precisely the same and reports failures. In the present application, where $0 < u(t) \leq 0.8$, we expect round-off errors to be of size 10^{-16}, so comparing 15 digits would probably be reliable, but we compare 14 to be on the safe side. On the other hand, comparing a small number of digits may hide software errors.

Doctests are highly encouraged as they do two things: 1) demonstrate how a function is used and 2) test that the function works.

5.3.2 Unit Tests and Test Functions

The unit testing technique consists of identifying smaller units of code and writing one or more tests for each unit. One unit can typically be a function. Each test should, ideally, not depend on the outcome of other tests. The recommended practice is actually to design and write the unit tests first and *then* implement the functions!

In scientific computing it is not always obvious how to best perform unit testing. The units are naturally larger than in non-scientific software. Very often the solution procedure of a mathematical problem identifies a unit, such as our `solver` function.

Two Python test frameworks: nose and pytest Python offers two very easy-to-use software frameworks for implementing unit tests: nose and pytest. These work (almost) in the same way, but our recommendation is to go for pytest.

Test function requirements For a test to qualify as a *test function* in nose or pytest, three rules must be followed:

1. The function name must start with `test_`.
2. Function arguments are not allowed.
3. An `AssertionError` exception must be raised if the test fails.

A specific example might be illustrative before proceeding. We have the following function that we want to test:

```
def double(n):
    return 2*n
```

The corresponding test function could, in principle, have been written as

```
def test_double():
    """Test that double(n) works for one specific n."""
    n = 4
    expected = 2*4
    computed = double(4)
    if expected != computed:
        raise AssertionError
```

The last two lines, however, are never written like this in test functions. Instead, Python's `assert` statement is used: `assert success, msg`, where `success` is a boolean variable, which is `False` if the test fails, and `msg` is *an optional* message string that is printed when the test fails. A better version of the test function is therefore

```
def test_double():
    """Test that double(n) works for one specific n."""
    n = 4
    expected = 2*4
    computed = double(4)
    msg = 'expected %g, computed %g' % (expected, computed)
    success = expected == computed
    assert success, msg
```

Comparison of real numbers Because of the finite precision arithmetics on a computer, which gives rise to round-off errors, the == operator is not suitable for checking whether two real numbers are equal. Obviously, this principle also applies to tests in test functions. We must therefore replace `a == b` by a comparison based on a tolerance `tol`: `abs(a-b) < tol`. The next example illustrates the problem and its solution.

Here is a slightly different function that we want to test:

```
def third(x):
    return x/3.
```

We write a test function where the expected result is computed as $\frac{1}{3}x$ rather than $x/3$:

```
def test_third():
    """Check that third(x) works for many x values."""
    for x in np.linspace(0, 1, 21):
        expected = (1/3.0)*x
        computed = third(x)
        success = expected == computed
        assert success
```

This `test_third` function executes silently, i.e., no failure, until `x` becomes 0.15. Then round-off errors make the == comparison `False`. In fact, seven of the `x` values above face this problem. The solution is to compare `expected` and `computed` with a small tolerance:

```
def test_third():
    """Check that third(x) works for many x values."""
    for x in np.linspace(0, 1, 21):
        expected = (1/3.)*x
        computed = third(x)
        tol = 1E-15
        success = abs(expected - computed) < tol
        assert success
```

Always compare real numbers with a tolerance!

Real numbers should never be compared with the == operator, but always with the absolute value of the difference and a tolerance. So, replace `a == b`, if `a` and/or `b` is `float`, by

```
tol = 1E-14
abs(a - b) < tol
```

The suitable size of `tol` depends on the size of `a` and `b` (see Problem 5.5). Unless `a` and `b` are around unity in size, one should use a *relative difference*:

```
tol = 1E-14
abs((a - b)/a) < tol
```

Special assert functions from nose Test frameworks often contain more tailored *assert functions* that can be called instead of using the `assert` statement. For example, comparing two objects within a tolerance, as in the present case, can be done by the `assert_almost_equal` from the nose framework:

```
import nose.tools as nt

def test_third():
    x = 0.15
    expected = (1/3.)*x
    computed = third(x)
    nt.assert_almost_equal(
        expected, computed, delta=1E-15,
        msg='diff=%.17E' % (expected - computed))
```

Whether to use the plain `assert` statement with a comparison based on a tolerance or to use the ready-made function `assert_almost_equal` depends on the programmer's preference. The examples used in the documentation of the pytest framework stick to the plain `assert` statement.

Locating test functions Test functions can reside in a module together with the functions they are supposed to verify, or the test functions can be collected in separate files having names starting with `test`. Actually, nose and pytest can recursively run all test functions in all `test*.py` files in the current directory, as well as in all subdirectories!

The decay.py[12] module file features test functions in the module, but we could equally well have made a subdirectory tests and put the test functions in tests/test_decay.py[13].

Running tests To run all test functions in the file decay.py do

```
Terminal> nosetests -s -v decay.py
Terminal> py.test -s -v decay.py
```

The -s option ensures that output from the test functions is printed in the terminal window, while -v prints the outcome of each individual test function.

Alternatively, if the test functions are located in some separate test*.py files, we can just write

```
Terminal> py.test -s -v
```

to *recursively* run *all* test functions in the current directory tree. The corresponding

```
Terminal> nosetests -s -v
```

command does the same, but requires subdirectory names to start with test or end with _test or _tests (which is a good habit anyway). An example of a tests directory with a test*.py file is found in src/softeng/tests[14].

Embedding doctests in a test function Doctests can also be executed from nose/pytest unit tests. Here is an example of a file test_decay_doctest.py[15] where we in the test block run all the doctests in the imported module decay, but we also include a local test function that does the same:

```
import sys, os
sys.path.insert(0, os.pardir)
import decay
import doctest

def test_decay_module_with_doctest():
    """Doctest embedded in a nose/pytest unit test."""
    # Test all functions with doctest in module decay
    failure_count, test_count = doctest.testmod(m=decay)
    assert failure_count == 0

if __name__ == '__main__':
    # Run all functions with doctests in this module
    failure_count, test_count = doctest.testmod(m=decay)
```

[12] http://tinyurl.com/ofkw6kc/softeng/decay.py
[13] http://tinyurl.com/ofkw6kc/softeng/tests/test_decay.py
[14] http://tinyurl.com/ofkw6kc/softeng/tests
[15] http://tinyurl.com/ofkw6kc/softeng/tests/test_decay_doctest.py

Running this file as a program from the command line triggers the `doctest.testmod` call in the test block, while applying `py.test` or `nosetests` to the file triggers an import of the file and execution of the test function `test_decay_modue_with_doctest`.

Installing nose and pytest With `pip` available, it is trivial to install nose and/or pytest: `sudo pip install nose` and `sudo pip install pytest`.

5.3.3 Test Function for the Solver

Finding good test problems for verifying the implementation of numerical methods is a topic on its own. The challenge is that we very seldom know what the numerical errors are. For the present model problem (5.1)–(5.2) solved by (5.3) one can, fortunately, derive a formula for the numerical approximation:

$$u^n = I\left(\frac{1-(1-\theta)a\Delta t}{1+\theta a\Delta t}\right)^n .$$

Then we know that the implementation should produce numbers that agree with this formula to machine precision. The formula for u^n is known as an *exact discrete solution* of the problem and can be coded as

```
def u_discrete_exact(n, I, a, theta, dt):
    """Return exact discrete solution of the numerical schemes."""
    dt = float(dt)  # avoid integer division
    A = (1 - (1-theta)*a*dt)/(1 + theta*dt*a)
    return I*A**n
```

A test function can evaluate this solution on a time mesh and check that the `u` values produced by the `solver` function do not deviate with more than a small tolerance:

```
def test_u_discrete_exact():
    """Check that solver reproduces the exact discr. sol."""
    theta = 0.8; a = 2; I = 0.1; dt = 0.8
    Nt = int(8/dt)  # no of steps
    u, t = solver(I=I, a=a, T=Nt*dt, dt=dt, theta=theta)

    # Evaluate exact discrete solution on the mesh
    u_de = np.array([u_discrete_exact(n, I, a, theta, dt)
                     for n in range(Nt+1)])

    # Find largest deviation
    diff = np.abs(u_de - u).max()
    tol = 1E-14
    success = diff < tol
    assert success
```

Among important things to consider when constructing test functions is testing the effect of wrong input to the function being tested. In our `solver` function, for example, integer values of a, Δt, and θ may cause unintended integer division. We should therefore add a test to make sure our `solver` function does not fall into this potential trap:

```
def test_potential_integer_division():
    """Choose variables that can trigger integer division."""
    theta = 1; a = 1; I = 1; dt = 2
    Nt = 4
    u, t = solver(I=I, a=a, T=Nt*dt, dt=dt, theta=theta)
    u_de = np.array([u_discrete_exact(n, I, a, theta, dt)
                     for n in range(Nt+1)])
    diff = np.abs(u_de - u).max()
    assert diff < 1E-14
```

In more complicated problems where there is no exact solution of the numerical problem solved by the software, one must use the method of manufactured solutions, compute convergence rates for a series of Δt values, and check that the rates converges to the expected ones (from theory). This type of testing is fully explained in Sect. 3.1.6.

5.3.4 Test Function for Reading Positional Command-Line Arguments

The function `read_command_line_positional` extracts numbers from the command line. To test it, we must decide on a set of values for the input data, fill `sys.argv` accordingly, and check that we get the expected values:

```
def test_read_command_line_positional():
    # Decide on a data set of input parameters
    I = 1.6;  a = 1.8;  T = 2.2;  theta = 0.5
    dt_values = [0.1, 0.2, 0.05]
    # Expected return from read_command_line_positional
    expected = [I, a, T, theta, dt_values]
    # Construct corresponding sys.argv array
    sys.argv = [sys.argv[0], str(I), str(a), str(T), 'CN'] + \
               [str(dt) for dt in dt_values]
    computed = read_command_line_positional()
    for expected_arg, computed_arg in zip(expected, computed):
        assert expected_arg == computed_arg
```

Note that `sys.argv[0]` is always the program name and that we have to copy that string from the original `sys.argv` array to the new one we construct in the test function. (Actually, this test function destroys the original `sys.argv` that Python fetched from the command line.)

Any numerical code writer should always be skeptical to the use of the exact equality operator == in test functions, since round-off errors often come into play. Here, however, we set some real values, convert them to strings and convert back again to real numbers (of the same precision). This string-number conversion does not involve any finite precision arithmetics effects so we can safely use == in tests. Note also that the last element in `expected` and `computed` is the list `dt_values`, and == works for comparing two lists as well.

5.3.5 Test Function for Reading Option-Value Pairs

The function `read_command_line_argparse` can be verified with a test function that has the same setup as `test_read_command_line_positional` above. However, the construction of the command line is a bit more complicated. We find it convenient to construct the line as a string and then split the line into words to get the desired list `sys.argv`:

```
def test_read_command_line_argparse():
    I = 1.6;  a = 1.8;  T = 2.2;  theta = 0.5
    dt_values = [0.1, 0.2, 0.05]
    # Expected return from read_command_line_argparse
    expected = [I, a, T, theta, dt_values]
    # Construct corresponding sys.argv array
    command_line = '%s --a %s --I %s --T %s --scheme CN --dt ' % \
                   (sys.argv[0], a, I, T)
    command_line += ' '.join([str(dt) for dt in dt_values])
    sys.argv = command_line.split()
    computed = read_command_line_argparse()
    for expected_arg, computed_arg in zip(expected, computed):
        assert expected_arg == computed_arg
```

Recall that the Python function `zip` enables iteration over several lists, tuples, or arrays at the same time.

Let silent test functions speak up during development!

When you develop test functions in a module, it is common to use IPython for interactive experimentation:

```
In[1]: import decay

In[2]: decay.test_read_command_line_argparse()
```

Note that a working test function is completely silent! Many find it psychologically annoying to convince themselves that a completely silent function is doing the right things. It can therefore, during development of a test function, be convenient to insert print statements in the function to monitor that the function body is indeed executed. For example, one can print the expected and computed values in the terminal window:

```
def test_read_command_line_argparse():
    ...
    for expected_arg, computed_arg in zip(expected, computed):
        print expected_arg, computed_arg
        assert expected_arg == computed_arg
```

After performing this edit, we want to run the test again, but in IPython the module must first be reloaded (reimported):

```
In[3]: reload(decay)  # force new import

In[2]: decay.test_read_command_line_argparse()
1.6 1.6
1.8 1.8
2.2 2.2
0.5 0.5
[0.1, 0.2, 0.05] [0.1, 0.2, 0.05]
```

Now we clearly see the objects that are compared.

5.3.6 Classical Class-Based Unit Testing

The test functions written for the nose and pytest frameworks are very straightforward and to the point, with no framework-required boilerplate code. We just write the statements we need to get the computations and comparisons done, before applying the required `assert`.

The classical way of implementing unit tests (which derives from the JUnit object-oriented tool in Java) leads to much more comprehensive implementations with a lot of boilerplate code. Python comes with a built-in module `unittest` for doing this type of classical unit tests. Although nose or pytest are much more convenient to use than `unittest`, class-based unit testing in the style of `unittest` has a very strong position in computer science and is so widespread in the software industry that even computational scientists should have an idea how such unit test code is written. A short demo of `unittest` is therefore included next. (Readers who are not familiar with object-oriented programming in Python may find the text hard to understand, but one can safely jump to the next section.)

Suppose we have a function `double(x)` in a module file `mymod.py`:

```
def double(x):
    return 2*x
```

Unit testing with the aid of the `unittest` module consists of writing a file `test_mymod.py` for testing the functions in `mymod.py`. The individual tests must be methods with names starting with `test_` in a class derived from class `TestCase` in `unittest`. With one test method for the function `double`, the `test_mymod.py` file becomes

```
import unittest
import mymod

class TestMyCode(unittest.TestCase):
    def test_double(self):
        x = 4
        expected = 2*x
        computed = mymod.double(x)
        self.assertEqual(expected, computed)

if __name__ == '__main__':
    unittest.main()
```

The test is run by executing the test file test_mymod.py as a standard Python program. There is no support in unittest for automatically locating and running all tests in all test files in a directory tree.

We could use the basic assert statement as we did with nose and pytest functions, but those who write code based on unittest almost exclusively use the wide range of built-in assert functions such as assertEqual, assertNotEqual, assertAlmostEqual, to mention some of them.

Translation of the test functions from the previous sections to unittest means making a new file test_decay.py file with a test class TestDecay where the stand-alone functions for nose/pytest now become methods in this class.

```
import unittest
import decay
import numpy as np

def u_discrete_exact(n, I, a, theta, dt):
    ...

class TestDecay(unittest.TestCase):

    def test_exact_discrete_solution(self):
        theta = 0.8; a = 2; I = 0.1; dt = 0.8
        Nt = int(8/dt)  # no of steps
        u, t = decay.solver(I=I, a=a, T=Nt*dt, dt=dt, theta=theta)
        # Evaluate exact discrete solution on the mesh
        u_de = np.array([u_discrete_exact(n, I, a, theta, dt)
                         for n in range(Nt+1)])
        diff = np.abs(u_de - u).max()  # largest deviation
        self.assertAlmostEqual(diff, 0, delta=1E-14)

    def test_potential_integer_division(self):
        ...
        self.assertAlmostEqual(diff, 0, delta=1E-14)

    def test_read_command_line_positional(self):
        ...
        for expected_arg, computed_arg in zip(expected, computed):
            self.assertEqual(expected_arg, computed_arg)

    def test_read_command_line_argparse(self):
        ...

if __name__ == '__main__':
    unittest.main()
```

5.4 Sharing the Software with Other Users

As soon as you have some working software that you intend to share with others, you should package your software in a standard way such that users can easily download your software, install it, improve it, and ask you to approve their improvements in new versions of the software. During recent years, the software development community has established quite firm tools and rules for how all this is done. The following subsections cover three steps in sharing software:

1. Organizing the software for public distribution.
2. Uploading the software to a cloud service (here GitHub).
3. Downloading and installing the software.

5.4.1 Organizing the Software Directory Tree

We start with organizing our software as a directory tree. Our software consists of one module file, `decay.py`, and possibly some unit tests in a separate file located in a directory `tests`.

The `decay.py` can be used as a module or as a program. For distribution to other users who install the program `decay.py` in system directories, we need to insert the following line at the top of the file:

```
#!/usr/bin/env python
```

This line makes it possible to write just the filename and get the file executed by the `python` program (or more precisely, the first `python` program found in the directories in the `PATH` environment variable).

Distributing just a module file Let us start out with the minimum solution alternative: distributing just the `decay.py` file. Then the software is just one file and all we need is a directory with this file. This directory will also contain an installation script `setup.py` and a `README` file telling what the software is about, the author's email address, a URL for downloading the software, and other useful information.

The `setup.py` file can be as short as

```
from distutils.core import setup
setup(name='decay',
      version='0.1',
      py_modules=['decay'],
      scripts=['decay.py'],
      )
```

The `py_modules` argument specifies a list of modules to be installed, while `scripts` specifies stand-alone programs. Our `decay.py` can be used either as a module or as an executable program, so we want users to have both possibilities.

Distributing a package If the software consists of more files than one or two modules, one should make a Python *package* out of it. In our case we make a package `decay` containing one module, also called `decay`.

To make a package `decay`, create a directory `decay` and an empty file in it with name `__init__.py`. A `setup.py` script must now specify the directory name of the package and also an executable program (`scripts=`) in case we want to run `decay.py` as a stand-alone application:

```
from distutils.core import setup
import os

setup(name='decay',
      version='0.1',
      author='Hans Petter Langtangen',
      author_email='hpl@simula.no',
      url='https://github.com/hplgit/decay-package/',
      packages=['decay'],
      scripts=[os.path.join('decay', 'decay.py')]
     )
```

We have also added some author and download information. The reader is referred to the Distutils documentation[16] for more information on how to write `setup.py` scripts.

Remark about the executable file

The executable program, `decay.py`, is in the above installation script taken to be the complete module file `decay.py`. It would normally be preferred to instead write a very short script essentially importing `decay` and running the test block in `decay.py`. In this way, we distribute a module and a very short file, say `decay-main.py`, as an executable program:

```
#!/usr/bin/env python
import decay
decay.decay.experiment_compare_dt(True)
decay.decay.plt.show()
```

In this package example, we move the unit tests out of the `decay.py` module to a separate file, `test_decay.py`, and place this file in a directory `tests`. Then the `nosetests` and `py.test` programs will automatically find and execute the tests.

The complete directory structure reads

```
Terminal

Terminal> /bin/ls -R
.:
decay  README  setup.py

./decay:
decay.py  __init__.py  tests

./decay/tests:
test_decay.py
```

5.4.2 Publishing the Software at GitHub

The leading site today for publishing open source software projects is GitHub at http://github.com, provided you want your software to be open to the world. With a paid GitHub account, you can have private projects too.

Sign up for a GitHub account if you do not already have one. Go to your account settings and provide an SSH key (typically the file `~/.ssh/id_rsa.pub`) such that you can communicate with GitHub without being prompted for your password. All communication between your computer and GitHub goes via the version control system Git. This may at first sight look tedious, but this is the way professionals work with software today. With Git you have full control of the history of your files, i.e., "who did what when". The technology makes Git superior to simpler alternatives like Dropbox and Google Drive, especially when you collaborate with others. There is a reason why Git has gained the position it has, and there is no reason why you should not adopt this tool.

[16] https://docs.python.org/2/distutils/setupscript.html

To create a new project, click on *New repository* on the main page and fill out a project name. Click on the check button *Initialize this repository with a README*, and click on *Create repository*. The next step is to clone (copy) the GitHub repo (short for repository) to your own computer(s) and fill it with files. The typical clone command is

Terminal

```
Terminal> git clone git://github.com:username/projname.git
```

where `username` is your GitHub username and `projname` is the name of the repo (project). The result of `git clone` is a directory `projname`. Go to this directory and add files. As soon as the repo directory is populated with files, run

Terminal

```
Terminal> git add .
Terminal> git commit -am 'First registration of project files'
Terminal> git push origin master
```

The above `git` commands look cryptic, but these commands plus 2–3 more are the essence of what you need in your daily work with files in small or big software projects. I strongly encourage you to learn more about version control systems and project hosting sites[17] [6].

Your project files are now stored in the cloud at https://github.com/username/projname. Anyone can get the software by the listed `git clone` command you used above, or by clicking on the links for zip and tar files.

Every time you update the project files, you need to register the update at GitHub by

Terminal

```
Terminal> git commit -am 'Description of the changes you made...'
Terminal> git push origin master
```

The files at GitHub are now synchronized with your local ones. Similarly, every time you start working on files in this project, make sure you have the latest version: `git pull origin master`.

You are recommended to read a quick intro[18] that makes you up and going with this style of professional work. And you should put all your writings and programming projects in repositories in the cloud!

5.4.3 Downloading and Installing the Software

Users of your software go to the Git repo at `github.com` and clone the repository. One can use either SSH or HTTP for communication. Most users will use the latter, typically

[17] http://hplgit.github.io/teamods/bitgit/html/
[18] http://hplgit.github.io/teamods/bitgit/html/

Terminal

```
Terminal> git clone https://github.com/username/projname.git
```

The result is a directory `projname` with the files in the repo.

Installing just a module file The software package is in the case above a directory `decay` with three files

Terminal

```
Terminal> ls decay
README    decay.py    setup.py
```

To install the `decay.py` file, a user just runs `setup.py`:

Terminal

```
Terminal> sudo python setup.py install
```

This command will install the software in system directories, so the user needs to run the command as `root` on Unix systems (therefore the command starts with `sudo`). The user can now import the module by `import decay` and run the program by

Terminal

```
Terminal> decay.py
```

A user can easily install the software on her personal account if a system-wide installation is not desirable. We refer to the installation documentation[19] for the many arguments that can be given to `setup.py`. Note that if the software is installed on a personal account, the `PATH` and `PYTHONPATH` environment variables must contain the relevant directories.

Our `setup.py` file specifies a module `decay` to be installed as well as a program `decay.py`. Modules are typically installed in some `lib` directory on the computer system, e.g., `/usr/local/lib/python2.7/dist-packages`, while executable programs go to `/usr/local/bin`.

Installing a package When the software is organized as a Python package, the installation is done by running `setup.py` exactly as explained above, but the use of a module `decay` in a package `decay` requires the following syntax:

```
import decay
u, t = decay.decay.solver(...)
```

That is, the call goes like `packagename.modulename.functionname`.

[19] https://docs.python.org/2/install/index.html#alternate-installation

Package import in `__init__.py`

One can ease the use of packages by providing a somewhat simpler import like

```
import decay
u, t = decay.solver(...)

# or
from decay import solver
u, t = solver(...)
```

This is accomplished by putting an import statement in the `__init__.py` file, which is always run when doing the package import `import decay` or `from decay import`. The `__init__.py` file must now contain

```
from decay import *
```

Obviously, it is the package developer who decides on such an `__init__.py` file or if it should just be empty.

5.5 Classes for Problem and Solution Method

The numerical solution procedure was compactly and conveniently implemented in a Python function `solver` in Sect. 5.1.1. In more complicated problems it might be beneficial to use classes instead of functions only. Here we shall describe a class-based software design well suited for scientific problems where there is a mathematical model of some physical phenomenon, and some numerical methods to solve the equations involved in the model.

We introduce a class `Problem` to hold the definition of the physical problem, and a class `Solver` to hold the data and methods needed to numerically solve the problem. The forthcoming text will explain the inner workings of these classes and how they represent an alternative to the `solver` and `experiment_*` functions in the `decay` module.

Explaining the details of class programming in Python is considered far beyond the scope of this text. Readers who are unfamiliar with Python class programming should first consult one of the many electronic Python tutorials or textbooks to come up to speed with concepts and syntax of Python classes before reading on. The author has a gentle introduction to class programming for scientific applications in [8], see Chapter 7 and 9 and Appendix E[20]. Other useful resources are

- The Python Tutorial: http://docs.python.org/2/tutorial/classes.html
- Wiki book on Python Programming: http://en.wikibooks.org/wiki/Python_Programming/Classes
- `tutorialspoint.com`: http://www.tutorialspoint.com/python/python_classes_objects.htm[21]

[20] http://hplgit.github.io/primer.html/doc/web/index.html

[21] http://www.tutorialspoint.com/python/python_classes_objects.htm

5.5.1 The Problem Class

The purpose of the problem class is to store all information about the mathematical model. This usually means the physical parameters and formulas in the problem. Looking at our model problem (5.1)–(5.2), the physical data cover I, a, and T. Since we have an analytical solution of the ODE problem, we may add this solution in terms of a Python function (or method) to the problem class as well. A possible problem class is therefore

```
from numpy import exp

class Problem(object):
    def __init__(self, I=1, a=1, T=10):
        self.T, self.I, self.a = I, float(a), T

    def u_exact(self, t):
        I, a = self.I, self.a
        return I*exp(-a*t)
```

We could in the u_exact method have written self.I*exp(-self.a*t), but using local variables I and a allows the nicer formula I*exp(-a*t), which looks much closer to the mathematical expression Ie^{-at}. This is not an important issue with the current compact formula, but is beneficial in more complicated problems with longer formulas to obtain the closest possible relationship between code and mathematics. The coding style in this book is to strip off the self prefix when the code expresses mathematical formulas.

The class data can be set either as arguments in the constructor or at any time later, e.g.,

```
problem = Problem(T=5)
problem.T = 8
problem.dt = 1.5
```

(Some programmers prefer set and get functions for setting and getting data in classes, often implemented via *properties* in Python, but this author considers that overkill when there are just a few data items in a class.)

It would be convenient if class Problem could also initialize the data from the command line. To this end, we add a method for defining a set of command-line options and a method that sets the local attributes equal to what was found on the command line. The default values associated with the command-line options are taken as the values provided to the constructor. Class Problem now becomes

```
class Problem(object):
    def __init__(self, I=1, a=1, T=10):
        self.T, self.I, self.a = I, float(a), T

    def define_command_line_options(self, parser=None):
        """Return updated (parser) or new ArgumentParser object."""
        if parser is None:
            import argparse
            parser = argparse.ArgumentParser()
```

```
        parser.add_argument(
            '--I', '--initial_condition', type=float,
            default=1.0, help='initial condition, u(0)',
            metavar='I')
        parser.add_argument(
            '--a', type=float, default=1.0,
            help='coefficient in ODE', metavar='a')
        parser.add_argument(
            '--T', '--stop_time', type=float,
            default=1.0, help='end time of simulation',
            metavar='T')
        return parser

    def init_from_command_line(self, args):
        """Load attributes from ArgumentParser into instance."""
        self.I, self.a, self.T = args.I, args.a, args.T

    def u_exact(self, t):
        """Return the exact solution u(t)=I*exp(-a*t)."""
        I, a = self.I, self.a
        return I*exp(-a*t)
```

Observe that if the user already has an `ArgumentParser` object it can be supplied, but if she does not have any, class `Problem` makes one. Python's `None` object is used to indicate that a variable is not initialized with a proper value.

5.5.2 The Solver Class

The solver class stores parameters related to the numerical solution method and provides a function `solve` for solving the problem. For convenience, a problem object is given to the constructor in a solver object such that the object gets access to the physical data. In the present example, the numerical solution method involves the parameters Δt and θ, which then constitute the data part of the solver class. We include, as in the problem class, functionality for reading Δt and θ from the command line:

```
class Solver(object):
    def __init__(self, problem, dt=0.1, theta=0.5):
        self.problem = problem
        self.dt, self.theta = float(dt), theta

    def define_command_line_options(self, parser):
        """Return updated (parser) or new ArgumentParser object."""
        parser.add_argument(
            '--scheme', type=str, default='CN',
            help='FE, BE, or CN')
        parser.add_argument(
            '--dt', '--time_step_values', type=float,
            default=[1.0], help='time step values',
            metavar='dt', nargs='+', dest='dt_values')
        return parser

    def init_from_command_line(self, args):
        """Load attributes from ArgumentParser into instance."""
        self.dt, self.theta = args.dt, args.theta
```

```
    def solve(self):
        self.u, self.t = solver(
            self.problem.I, self.problem.a, self.problem.T,
            self.dt, self.theta)

    def error(self):
        """Return norm of error at the mesh points."""
        u_e = self.problem.u_exact(self.t)
        e = u_e - self.u
        E = np.sqrt(self.dt*np.sum(e**2))
        return E
```

Note that we see no need to repeat the body of the previously developed and tested `solver` function. We just call that function from the `solve` method. In this way, class `Solver` is merely a class wrapper of the stand-alone `solver` function. With a single object of class `Solver` we have all the physical and numerical data bundled together with the numerical solution method.

Combining the objects Eventually we need to show how the classes `Problem` and `Solver` play together. We read parameters from the command line and make a plot with the numerical and exact solution:

```
def experiment_classes():
    problem = Problem()
    solver = Solver(problem)

    # Read input from the command line
    parser = problem.define_command_line_options()
    parser = solver. define_command_line_options(parser)
    args = parser.parse_args()
    problem.init_from_command_line(args)
    solver. init_from_command_line(args)

    # Solve and plot
    solver.solve()
    import matplotlib.pyplot as plt
    t_e = np.linspace(0, T, 1001)    # very fine mesh for u_e
    u_e = problem.u_exact(t_e)
    print 'Error:', solver.error()

    plt.plot(t,   u,   'r--o')
    plt.plot(t_e, u_e, 'b-')
    plt.legend(['numerical, theta=%g' % theta, 'exact'])
    plt.xlabel('t')
    plt.ylabel('u')
    plotfile = 'tmp'
    plt.savefig(plotfile + '.png');  plt.savefig(plotfile + '.pdf')
    plt.show()
```

5.5.3 Improving the Problem and Solver Classes

The previous `Problem` and `Solver` classes containing parameters soon get much repetitive code when the number of parameters increases. Much of this code can be parameterized and be made more compact. For this purpose, we decide to collect all parameters in a dictionary, `self.prm`, with two associated dictionaries `self.type` and `self.help` for holding associated object types and help strings. The reason is that processing dictionaries is easier than processing a set of individual attributes.

For the specific ODE example we deal with, the three dictionaries in the problem class are typically

```
self.prm  = dict(I=1, a=1, T=10)
self.type = dict(I=float, a=float, T=float)
self.help = dict(I='initial condition, u(0)',
                 a='coefficient in ODE',
                 T='end time of simulation')
```

Provided a problem or solver class defines these three dictionaries in the constructor, we can create a super class `Parameters` with general code for defining command-line options and reading them as well as methods for setting and getting each parameter. A `Problem` or `Solver` for a particular mathematical problem can then inherit most of the needed functionality and code from the `Parameters` class. For example,

```
class Problem(Parameters):
    def __init__(self):
        self.prm  = dict(I=1, a=1, T=10)
        self.type = dict(I=float, a=float, T=float)
        self.help = dict(I='initial condition, u(0)',
                         a='coefficient in ODE',
                         T='end time of simulation')

    def u_exact(self, t):
        I, a = self['I a'.split()]
        return I*np.exp(-a*t)

class Solver(Parameters):
    def __init__(self, problem):
        self.problem = problem   # class Problem object
        self.prm  = dict(dt=0.5, theta=0.5)
        self.type = dict(dt=float, theta=float)
        self.help = dict(dt='time step value',
                         theta='time discretization parameter')

    def solve(self):
        from decay import solver
        I, a, T = self.problem['I a T'.split()]
        dt, theta = self['dt theta'.split()]
        self.u, self.t = solver(I, a, T, dt, theta)
```

By inheritance, these classes can automatically do a lot more when it comes to reading and assigning parameter values:

```
problem = Problem()
solver = Solver(problem)

# Read input from the command line
parser = problem.define_command_line_options()
parser = solver. define_command_line_options(parser)
args = parser.parse_args()
problem.init_from_command_line(args)
solver. init_from_command_line(args)

# Other syntax for setting/getting parameter values
problem['T'] = 6
print 'Time step:', solver['dt']

solver.solve()
u, t = solver.u, solver.t
```

A generic class for parameters A simplified version of the parameter class looks as follows:

```
class Parameters(object):
    def __getitem__(self, name):
        """obj[name] syntax for getting parameters."""
        if isinstance(name, (list,tuple)):          # get many?
            return [self.prm[n] for n in name]
        else:
            return self.prm[name]

    def __setitem__(self, name, value):
        """obj[name] = value syntax for setting a parameter."""
        self.prm[name] = value

    def define_command_line_options(self, parser=None):
        """Automatic registering of options."""
        if parser is None:
            import argparse
            parser = argparse.ArgumentParser()

        for name in self.prm:
            tp = self.type[name] if name in self.type else str
            help = self.help[name] if name in self.help else None
            parser.add_argument(
                '--' + name, default=self.get(name), metavar=name,
                type=tp, help=help)

        return parser

    def init_from_command_line(self, args):
        for name in self.prm:
            self.prm[name] = getattr(args, name)
```

The file `decay_oo.py`[22] contains a slightly more advanced version of class `Parameters` where the functions for getting and setting parameters contain tests for valid parameter names, and raise exceptions with informative messages if any name is not registered.

We have already sketched the `Problem` and `Solver` classes that build on inheritance from `Parameters`. We have also shown how they are used. The only remaining code is to make the plot, but this code is identical to previous versions when the numerical solution is available in an object u and the exact one in `u_e`.

The advantage with the `Parameters` class is that it scales to problems with a large number of physical and numerical parameters: as long as the parameters are defined once via a dictionary, the compact code in class `Parameters` can handle any collection of parameters of any size. More advanced tools for storing large collections of parameters in hierarchical structures is provided by the Parampool[23] package.

[22] http://tinyurl.com/ofkw6kc/softeng/decay_oo.py
[23] https://github.com/hplgit/parampool

5.6 Automating Scientific Experiments

Empirical scientific investigations based on running computer programs require careful design of the experiments and accurate reporting of results. Although there is a strong tradition to do such investigations manually, the extreme requirements to scientific accuracy make a program much better suited to conduct the experiments. We shall in this section outline how we can write such programs, often called *scripts*, for running other programs and archiving the results.

Scientific investigation

The purpose of the investigations is to explore the quality of numerical solutions to an ordinary differential equation. More specifically, we solve the initial-value problem

$$u'(t) = -au(t), \quad u(0) = I, \quad t \in (0, T], \tag{5.4}$$

by the θ-rule:

$$u^{n+1} = \frac{1 - (1 - \theta)a\Delta t}{1 + \theta a \Delta t} u^n, \quad u^0 = I\,. \tag{5.5}$$

This scheme corresponds to well-known methods: $\theta = 0$ gives the Forward Euler (FE) scheme, $\theta = 1$ gives the Backward Euler (BE) scheme, and $\theta = \frac{1}{2}$ gives the Crank–Nicolson (CN) or midpoint/centered scheme.

For chosen constants I, a, and T, we run the three schemes for various values of Δt, and present the following results in a report:

1. visual comparison of the numerical and exact solution in a plot for each Δt and $\theta = 0, 1, \frac{1}{2}$,
2. a table and a plot of the norm of the numerical error versus Δt for $\theta = 0, 1, \frac{1}{2}$.

The report will also document the mathematical details of the problem under investigation.

5.6.1 Available Software

Appropriate software for implementing (5.5) is available in a program `model.py`[24], which is run as

```
Terminal
Terminal> python model.py --I 1.5 --a 0.25 --T 6 --dt 1.25 0.75 0.5
```

The command-line input corresponds to setting $I = 1.5$, $a = 0.25$, $T = 6$, and run three values of Δt: 1.25, 0.75, ad 0.5.

The results of running this `model.py` command are text in the terminal window and a set of plot files. The plot files have names `M_D.E`, where `M` denotes the method (`FE`, `BE`, `CN` for $\theta = 0, 1, \frac{1}{2}$, respectively), `D` the time step length (here `1.25`, `0.75`,

[24] http://tinyurl.com/nc4upel/doconce_src/model.py

or `0.5`), and `E` is the plot file extension `png` or `pdf`. The text output in the terminal window looks like

```
0.0   1.25:    5.998E-01
0.0   0.75:    1.926E-01
0.0   0.50:    1.123E-01
0.0   0.10:    1.558E-02
0.5   1.25:    6.231E-02
0.5   0.75:    1.543E-02
0.5   0.50:    7.237E-03
0.5   0.10:    2.469E-04
1.0   1.25:    1.766E-01
1.0   0.75:    8.579E-02
1.0   0.50:    6.884E-02
1.0   0.10:    1.411E-02
```

The first column is the θ value, the next the Δt value, and the final column represents the numerical error E (the norm of discrete error function on the mesh).

5.6.2 The Results We Want to Produce

The results we need for our investigations are slightly different than what is directly produced by `model.py`:

1. We need to collect all the plots for one numerical method (FE, BE, CN) in a single plot. For example, if 4 Δt values are run, the summarizing figure for the BE method has 2×2 subplots, with the subplot corresponding to the largest Δt in the upper left corner and the smallest in the bottom right corner.
2. We need to create a table containing Δt values in the first column and the numerical error E for $\theta = 0, 0.5, 1$ in the next three columns. This table should be available as a standard CSV file.
3. We need to plot the numerical error E versus Δt in a log-log plot.

Consequently, we must write a script that can run `model.py` as described and produce the results 1–3 above. This requires combining multiple plot files into one file and interpreting the output from `model.py` as data for plotting and file storage.

If the script's name is `exper1.py`, we run it with the desired Δt values as positional command-line arguments:

Terminal
```
Terminal> python exper1.py 0.5 0.25 0.1 0.05
```

This run will then generate eight plot files: `FE.png` and `FE.pdf` summarizing the plots with the FE method, `BE.png` and `BE.pdf` with the BE method, `CN.png` and `CN.pdf` with the CN method, and `error.png` and `error.pdf` with the log-log plot of the numerical error versus Δt. In addition, the table with numerical errors is written to a file `error.csv`.

Reproducible and replicable science

A script that automates running our computer experiments will ensure that the experiments can easily be rerun by anyone in the future, either to confirm the same results or redo the experiments with other input data. Also, whatever we did to produce the results is documented in every detail in the script.

A project where anyone can easily repeat the experiments with the same data is referred to as being *replicable*, and replicability should be a fundamental requirement in scientific computing work. Of more scientific interest is *reproducibilty*, which means that we can also run alternative experiments to arrive at the same conclusions. This requires more than an automating script.

5.6.3 Combining Plot Files

The script for running experiments needs to combine multiple image files into one. The `montage`[25] and `convert`[26] programs in the ImageMagick software suite can be used to combine image files. However, these programs are best suited for PNG files. For vector plots in PDF format one needs other tools to preserve the quality: `pdftk`, `pdfnup`, and `pdfcrop`.

Suppose you have four files `f1.png`, `f2.png`, `f3.png`, and `f4.png` and want to combine them into a 2×2 table of subplots in a new file `f.png`, see Fig. 5.4 for an example.

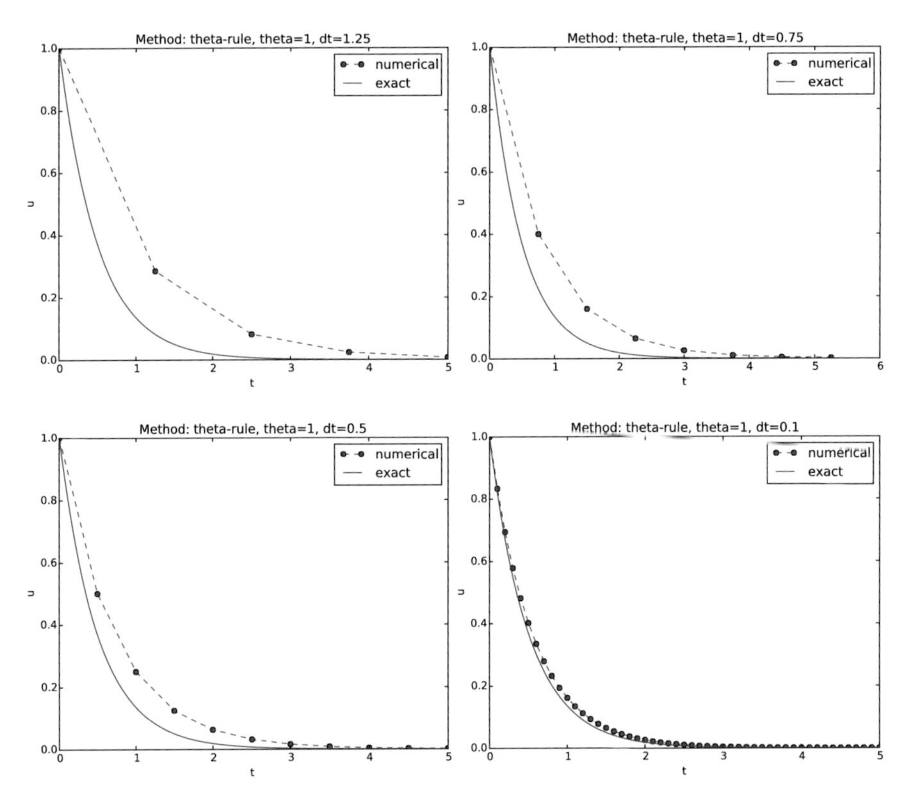

Fig. 5.4 Illustration of the Backward Euler method for four time step values

[25] http://www.imagemagick.org/script/montage.php
[26] http://www.imagemagick.org/script/convert.php

The appropriate ImageMagick commands are

```
Terminal

Terminal> montage -background white -geometry 100% -tile 2x \
          f1.png f2.png f3.png f4.png f.png
Terminal> convert -trim f.png f.png
Terminal> convert f.png -transparent white f.png
```

The first command mounts the four files in one, the next `convert` command removes unnecessary surrounding white space, and the final `convert` command makes the white background transparent.

High-quality montage of PDF files `f1.pdf`, `f2.pdf`, `f3.pdf`, and `f4.pdf` into `f.pdf` goes like

```
Terminal

Terminal> pdftk f1.pdf f2.pdf f3.pdf f4.pdf output tmp.pdf
Terminal> pdfnup --nup 2x2 --outfile tmp.pdf tmp.pdf
Terminal> pdfcrop tmp.pdf f.pdf
Terminal> rm -f tmp.pdf
```

5.6.4 Running a Program from Python

The script for automating experiments needs to run the `model.py` program with appropriate command-line options. Python has several tools for executing an arbitrary command in the operating systems. Let `cmd` be a string containing the desired command. In the present case study, `cmd` could be `'python model.py -I 1 -dt 0.5 0.2'`. The following code executes `cmd` and loads the text output into a string `output`:

```
from subprocess import Popen, PIPE, STDOUT
p = Popen(cmd, shell=True, stdout=PIPE, stderr=STDOUT)
output, _ = p.communicate()

# Check if the execution was successful
failure = p.returncode
if failure:
    print 'Command failed:', cmd; sys.exit(1)
```

Unsuccessful execution usually makes it meaningless to continue the program, and therefore we abort the program with `sys.exit(1)`. Any argument different from 0 signifies to the computer's operating system that our program stopped with a failure.

Programming tip: use _ for dummy variable

Sometimes we need to unpack tuples or lists in separate variables, but we are not interested in all the variables. One example is

```
output, error = p.communicate()
```

but `error` is of no interest in the example above. One can then use underscore _ as variable name for the dummy (uninteresting) variable(s):

```
output, _ = p.communicate()
```

Here is another example where we iterate over a list of three-tuples, but the interest is limited to the second element in each three-tuple:

```
for _, value, _ in list_of_three_tuples:
    # work with value
```

We need to interpret the contents of the string `output` and store the data in an appropriate data structure for further processing. Since the content is basically a table and will be transformed to a spread sheet format, we let the columns in the table be represented by lists in the program, and we collect these columns in a dictionary whose keys are natural column names: `dt` and the three values of θ. The following code translates the output of `cmd` (`output`) to such a dictionary of lists (`errors`):

```
errors = {'dt': dt_values, 1: [], 0: [], 0.5: []}
for line in output.splitlines():
    words = line.split()
    if words[0] in ('0.0', '0.5', '1.0'):  # line with E?
        # typical line: 0.0    1.25:     7.463E+00
        theta = float(words[0])
        E = float(words[2])
        errors[theta].append(E)
```

5.6.5 The Automating Script

We have now all the core elements in place to write the complete script where we run `model.py` for a set of Δt values (given as positional command-line arguments), make the error plot, write the CSV file, and combine plot files as described above. The complete code is listed below, followed by some explaining comments.

```
import os, sys, glob
import matplotlib.pyplot as plt

def run_experiments(I=1, a=2, T=5):
    # The command line must contain dt values
    if len(sys.argv) > 1:
        dt_values = [float(arg) for arg in sys.argv[1:]]
    else:
        print 'Usage: %s dt1 dt2 dt3 ...' %  sys.argv[0]
        sys.exit(1)  # abort

    # Run module file and grab output
    cmd = 'python model.py --I %g --a %g --T %g' % (I, a, T)
    dt_values_str = ' '.join([str(v) for v in dt_values])
    cmd += ' --dt %s' % dt_values_str
    print cmd
```

```
from subprocess import Popen, PIPE, STDOUT
p = Popen(cmd, shell=True, stdout=PIPE, stderr=STDOUT)
output, _ = p.communicate()
failure = p.returncode
if failure:
    print 'Command failed:', cmd; sys.exit(1)

errors = {'dt': dt_values, 1: [], 0: [], 0.5: []}
for line in output.splitlines():
    words = line.split()
    if words[0] in ('0.0', '0.5', '1.0'):  # line with E?
        # typical line: 0.0   1.25:    7.463E+00
        theta = float(words[0])
        E = float(words[2])
        errors[theta].append(E)

# Find min/max for the axis
E_min = 1E+20; E_max = -E_min
for theta in 0, 0.5, 1:
    E_min = min(E_min, min(errors[theta]))
    E_max = max(E_max, max(errors[theta]))

plt.loglog(errors['dt'], errors[0], 'ro-')
plt.loglog(errors['dt'], errors[0.5], 'b+-')
plt.loglog(errors['dt'], errors[1], 'gx-')
plt.legend(['FE', 'CN', 'BE'], loc='upper left')
plt.xlabel('log(time step)')
plt.ylabel('log(error)')
plt.axis([min(dt_values), max(dt_values), E_min, E_max])
plt.title('Error vs time step')
plt.savefig('error.png');  plt.savefig('error.pdf')

# Write out a table in CSV format
f = open('error.csv', 'w')
f.write(r'$\Delta t$,$\theta=0$,$\theta=0.5$,$\theta=1$' \
        + '\n')
for _dt, _fe, _cn, _be in zip(
    errors['dt'], errors[0], errors[0.5], errors[1]):
    f.write('%.2f,%.4f,%.4f,%.4f\n' % \
            (_dt, _fe, _cn, _be))
f.close()

# Combine images into rows with 2 plots in each row
image_commands = []
for method in 'BE', 'CN', 'FE':
    pdf_files = ' '.join(['%s_%g.pdf' % (method, dt)
                          for dt in dt_values])
    png_files = ' '.join(['%s_%g.png' % (method, dt)
                          for dt in dt_values])
    image_commands.append(
        'montage -background white -geometry 100%' +
        ' -tile 2x %s %s.png' % (png_files, method))
    image_commands.append(
        'convert -trim %s.png %s.png' % (method, method))
    image_commands.append(
        'convert %s.png -transparent white %s.png' %
        (method, method))
    image_commands.append(
        'pdftk %s output tmp.pdf' % pdf_files)
    num_rows = int(round(len(dt_values)/2.0))
    image_commands.append(
        'pdfnup --nup 2x%d --outfile tmp.pdf tmp.pdf' % \
        num_rows)
    image_commands.append(
        'pdfcrop tmp.pdf %s.pdf' % method)
```

```
    for cmd in image_commands:
        print cmd
        failure = os.system(cmd)
        if failure:
            print 'Command failed:', cmd; sys.exit(1)

    # Remove the files generated above and by model.py
    from glob import glob
    filenames = glob('*_*.png') + glob('*_*.pdf') + \
                glob('tmp*.pdf')
    for filename in filenames:
        os.remove(filename)

if __name__ == '__main__':
    run_experiments(I=1, a=2, T=5)
    plt.show()
```

We may comment upon many useful constructs in this script:

- `[float(arg) for arg in sys.argv[1:]]` builds a list of real numbers from all the command-line arguments.
- `['%s_%s.png' % (method, dt) for dt in dt_values]` builds a list of filenames from a list of numbers (`dt_values`).
- All `montage`, `convert`, `pdftk`, `pdfnup`, and `pdfcrop` commands for creating composite figures are stored in a list and later executed in a loop.
- `glob('*_*.png')` returns a list of the names of all files in the current directory where the filename matches the Unix wildcard notation[27] `*_*.png` (meaning any text, underscore, any text, and then `.png`).
- `os.remove(filename)` removes the file with name `filename`.
- `failure = os.system(cmd)` runs an operating system command with simpler syntax than what is required by `subprocess` (but the output of `cmd` cannot be captured).

5.6.6 Making a Report

The results of running computer experiments are best documented in a little report containing the problem to be solved, key code segments, and the plots from a series of experiments. At least the part of the report containing the plots should be automatically generated by the script that performs the set of experiments, because in the script we know exactly which input data that were used to generate a specific plot, thereby ensuring that each figure is connected to the right data. Take a look at a sample report[28] to see what we have in mind.

Word, OpenOffice, GoogleDocs Microsoft Word, its open source counterparts OpenOffice and LibreOffice, along with GoogleDocs and similar online services are the dominating tools for writing reports today. Nevertheless, scientific reports often need mathematical equations and nicely typeset computer code in monospace font. The support for mathematics and computer code in the mentioned tools is

[27] http://en.wikipedia.org/wiki/Glob_(programming)
[28] http://tinyurl.com/nc4upel/_static/sphinx-cloud/

We address the initial-value problem

$$u'(t) = -au(t), \quad t \in (0,T], \tag{1}$$
$$u(0) = I, \tag{2}$$

where a, I, and T are prescribed parameters, and $u(t)$ is the unknown function to be estimated. This mathematical model is relevant for physical phenomena featuring exponential decay in time.

Numerical solution method

We introduce a mesh in time with points $0 = t_0 < t_1 \cdots < t_N = T$. For simplicity, we assume constant spacing Δt between the mesh points: $\Delta t = t_n - t_{n-1}$, $n = 1,\ldots,N$. Let u^n be the numerical approximation to the exact solution at t_n. The θ-rule is used to solve (1) numerically:

$$u^{n+1} = \frac{1-(1-\theta)a\Delta t}{1+\theta a\Delta t}u^n,$$

for $n = 0,1,\ldots,N-1$. This scheme corresponds to

- The Forward Euler scheme when $\theta = 0$
- The Backward Euler scheme when $\theta = 1$
- The Crank-Nicolson scheme when $\theta = 1/2$

Implementation

The numerical method is implemented in a Python function:

```
def theta_rule(I, a, T, dt, theta):
    """Solve u'=-a*u, u(0)=I, for t in (0,T] with steps of dt."""
    N = int(round(T/float(dt)))  # no of intervals
    u = zeros(N+1)
    t = linspace(0, T, N+1)
```

Fig. 5.5 Report in HTML format with MathJax

in this author's view not on par with the technologies based on *markup languages* and which are addressed below. Also, with markup languages one has a readable, pure text file as source for the report, and changes in this text can easily be tracked by version control systems like Git. The result is a very strong tool for monitoring "who did what when" with the files, resulting in increased reliability of the writing process. For collaborative writing, the merge functionality in Git leads to safer simultaneously editing than what is offered even by collaborative tools like GoogleDocs.

HTML with MathJax HTML is the markup language used for web pages. Nicely typeset computer code is straightforward in HTML, and high-quality mathematical typesetting is available using an extension to HTML called MathJax[29], which allows formulas and equations to be typeset with LaTeX syntax and nicely rendered in web browsers, see Fig. 5.5. A relatively small subset of LaTeX environments for mathematics is supported, but the syntax for formulas is quite rich. Inline formulas look like `\( u'=-au \)` while equations are surrounded by `$$` signs. Inside such signs, one can use `\[ u'=-au \]` for unnumbered equations, or `\begin{equation}` and `\end{equation}` for numbered equations, or `\begin{align}` and `\end{align}` for multiple numbered aligned equations. You need to be familiar with mathematical typesetting in LaTeX[30] to write MathJax code.

[29] http://www.mathjax.org/

[30] http://en.wikibooks.org/wiki/LaTeX/Mathematics

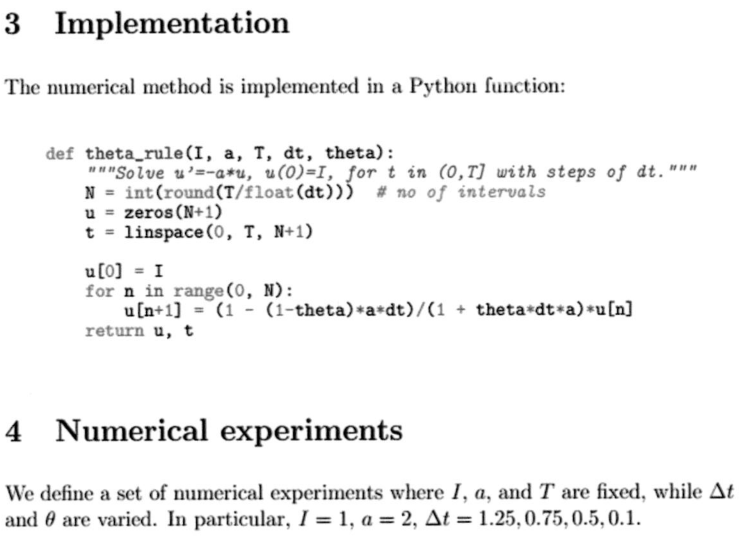

3 Implementation

The numerical method is implemented in a Python function:

```
def theta_rule(I, a, T, dt, theta):
    """Solve u'=-a*u, u(0)=I, for t in (0,T] with steps of dt."""
    N = int(round(T/float(dt)))  # no of intervals
    u = zeros(N+1)
    t = linspace(0, T, N+1)

    u[0] = I
    for n in range(0, N):
        u[n+1] = (1 - (1-theta)*a*dt)/(1 + theta*dt*a)*u[n]
    return u, t
```

4 Numerical experiments

We define a set of numerical experiments where I, a, and T are fixed, while Δt and θ are varied. In particular, $I = 1$, $a = 2$, $\Delta t = 1.25, 0.75, 0.5, 0.1$.

Fig. 5.6 Report in PDF format generated from LaTeX source

The file `exper1_mathjax.py`[31] calls a script `exper1.py`[32] to perform the numerical experiments and then runs Python statements for creating an HTML file[33] with the source code for the scientific report[34].

LaTeX The *de facto* language for mathematical typesetting and scientific report writing is LaTeX[35]. A number of very sophisticated packages have been added to the language over a period of three decades, allowing very fine-tuned layout and typesetting. For output in the PDF format[36], see Fig. 5.6 for an example, LaTeX is the definite choice when it comes to *typesetting quality*. The LaTeX language used to write the reports has typically a lot of commands involving backslashes and braces[37], and many claim that LaTeX syntax is not particularly readable. For output on the web via HTML code (i.e., not only showing the PDF in the browser window), LaTeX struggles with delivering high quality typesetting. Other tools, especially Sphinx, give better results and can also produce nice-looking PDFs. The file `exper1_latex.py`[38] shows how to generate the LaTeX source from a program.

Sphinx Sphinx[39] is a typesetting language with similarities to HTML and LaTeX, but with much less tagging. It has recently become very popular for software documentation and mathematical reports. Sphinx can utilize LaTeX for mathematical

[31] http://tinyurl.com/p96acy2/report_generation/exper1_html.py
[32] http://tinyurl.com/p96acy2/exper1.py
[33] http://tinyurl.com/nc4upel/_static/report_mathjax.html.html
[34] http://tinyurl.com/nc4upel/_static/report_mathjax.html
[35] http://en.wikipedia.org/wiki/LaTeX
[36] http://tinyurl.com/nc4upel/_static/report.pdf
[37] http://tinyurl.com/nc4upel/_static/report.tex.html
[38] http://tinyurl.com/p96acy2/report_generation/exper1_latex.py
[39] http://sphinx.pocoo.org/

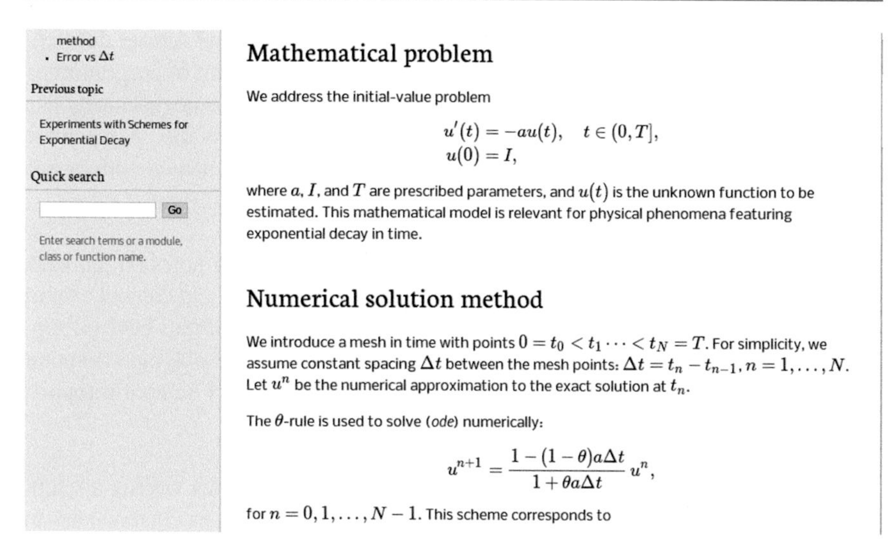

Fig. 5.7 Report in HTML format generated from Sphinx source

formulas and equations. Unfortunately, the subset of LaTeX mathematics supported is less than in full MathJax (in particular, numbering of multiple equations in an `align` type environment is not supported). The Sphinx syntax[40] is an extension of the reStructuredText language. An attractive feature of Sphinx is its rich support for fancy layout of web pages[41]. In particular, Sphinx can easily be combined with various layout *themes* that give a certain look and feel to the web site and that offers table of contents, navigation, and search facilities, see Fig. 5.7.

Markdown A recent, very popular format for easy writing of web pages is Markdown[42]. Text is written very much like one would do in email, using spacing and special characters to naturally format the code instead of heavily tagging the text as in LaTeX and HTML. With the tool Pandoc[43] one can go from Markdown to a variety of formats. HTML is a common output format, but LaTeX, epub, XML, OpenOffice/LibreOffice, MediaWiki, and Microsoft Word are some other possibilities. A Markdown version of our scientific report demo is available as an IPython/Jupyter notebook (described next).

IPython/Jupyter notebooks. The Jupyter Notebook[44] is a web-based tool where one can write scientific reports with live computer code and graphics. Or the other way around: software can be equipped with documentation in the style of scientific reports. A slightly extended version of Markdown is used for writing text and mathematics, and the source code of a notebook[45] is in json format. The interest in

[40] http://tinyurl.com/nc4upel/_static/report_sphinx.rst.html
[41] http://tinyurl.com/nc4upel/_static/sphinx-cloud/index.html
[42] http://daringfireball.net/projects/markdown/
[43] http://johnmacfarlane.net/pandoc/
[44] http://jupyter.org
[45] http://tinyurl.com/nc4upel/_static/report.ipynb.html

the notebook has grown amazingly fast over just a few years, and further development now takes place in the Jupyter project[46], which supports a lot of programming languages for interactive notebook computing. Jupyter notebooks are primarily live electronic documents, but they can be printed out as PDF reports too. A notebook version of our scientific report can be downloaded[47] and experimented with or just statically viewed[48] in a browser.

Wiki formats A range of wiki formats are popular for creating notes on the web, especially documents which allow groups of people to edit and add content. Apart from MediaWiki[49] (the wiki format used for Wikipedia), wiki formats have no support for mathematical typesetting and also limited tools for displaying computer code in nice ways. Wiki formats are therefore less suitable for scientific reports compared to the other formats mentioned here.

DocOnce Since it is difficult to choose the right tool or format for writing a scientific report, it is advantageous to write the content in a format that easily translates to LaTeX, HTML, Sphinx, Markdown, IPython/Jupyter notebooks, and various wikis. DocOnce[50] is such a tool. It is similar to Pandoc, but offers some special convenient features for writing about mathematics and programming. The tagging is modest[51], somewhere between LaTeX and Markdown. The program `exper1_do.py`[52] demonstrates how to generate DocOnce code for a scientific report. There is also a corresponding rich demo of the resulting reports[53] that can be made from this DocOnce code.

5.6.7 Publishing a Complete Project

To assist the important principle of *replicable* science, a report documenting scientific investigations should be accompanied by all the software and data used for the investigations so that others have a possibility to redo the work and assess the quality of the results.

One way of documenting a complete project is to make a directory tree with all relevant files. Preferably, the tree is published at some project hosting site like Bitbucket or GitHub[54] so that others can download it as a tarfile, zipfile, or fork the files directly using the Git version control system. For the investigations outlined in Sect. 5.6.6, we can create a directory tree with files

[46] https://jupyter.org/
[47] http://tinyurl.com/p96acy2/_static/report.ipynb
[48] http://nbviewer.ipython.org/url/hplgit.github.com/teamods/writing_reports/_static/report.ipynb
[49] http://www.mediawiki.org/wiki/MediaWiki
[50] https://github.com/hplgit/doconce
[51] http://tinyurl.com/nc4upel/_static/report.do.txt.html
[52] http://tinyurl.com/p96acy2/exper1_do.py
[53] http://tinyurl.com/nc4upel/index.html
[54] http://hplgit.github.com/teamods/bitgit/html/

```
setup.py
./src:
   model.py
./doc:
   ./src:
      exper1_mathjax.py
      make_report.sh
      run.sh
   ./pub:
      report.html
```

The `src` directory holds source code (modules) to be reused in other projects, the `setup.py` script builds and installs such software, the `doc` directory contains the documentation, with `src` for the source of the documentation (usually written in a markup language) and `pub` for published (compiled) documentation. The `run.sh` file is a simple Bash script listing the `python` commands we used to run `exper1_mathjax.py` to generate the experiments and the `report.html` file.

5.7 Exercises

Problem 5.1: Make a tool for differentiating curves
Suppose we have a curve specified through a set of discrete coordinates (x_i, y_i), $i = 0, \ldots, n$, where the x_i values are uniformly distributed with spacing Δx: $x_i = \Delta x$. The derivative of this curve, defined as a new curve with points (x_i, d_i), can be computed via finite differences:

$$d_0 = \frac{y_1 - y_0}{\Delta x}, \tag{5.6}$$

$$d_i = \frac{y_{i+1} - y_{i-1}}{2\Delta x}, \quad i = 1, \ldots, n-1, \tag{5.7}$$

$$d_n = \frac{y_n - y_{n-1}}{\Delta x}. \tag{5.8}$$

a) Write a function `differentiate(x, y)` for differentiating a curve with coordinates in the arrays `x` and `y`, using the formulas above. The function should return the coordinate arrays of the resulting differentiated curve.
b) Since the formulas for differentiation used here are only approximate, with unknown approximation errors, it is challenging to construct test cases. Here are three approaches, which should be implemented in three separate test functions.
   1. Consider a curve with three points and compute d_i, $i = 0, 1, 2$, by hand. Make a test that compares the hand-calculated results with those from the function in a).
   2. The formulas for d_i are exact for points on a straight line, as all the d_i values are then the same, equal to the slope of the line. A test can check this property.
   3. For points lying on a parabola, the values for d_i, $i = 1, \ldots, n-1$, should equal the exact derivative of the parabola. Make a test based on this property.
c) Start with a curve corresponding to $y = \sin(\pi x)$ and $n + 1$ points in $[0, 1]$. Apply `differentiate` four times and plot the resulting curve and the exact $y = \sin \pi x$ for $n = 6, 11, 21, 41$.

Filename: `curvediff`.

Problem 5.2: Make solid software for the Trapezoidal rule
An integral

$$\int_a^b f(x)dx$$

can be numerically approximated by the Trapezoidal rule,

$$\int_a^b f(x)dx \approx \frac{h}{2}(f(a)+f(b)) + h\sum_{i=1}^{n-1} f(x_i),$$

where x_i is a set of uniformly spaced points in $[a, b]$:

$$h = \frac{b-a}{n}, \quad x_i = a + ih,\ i = 1, \ldots, n-1\,.$$

Somebody has used this rule to compute the integral $\int_0^\pi \sin^2 x\, dx$:

```
from math import pi, sin
np = 20
h = pi/np
I = 0
for k in range(1, np):
    I += sin(k*h)**2
print I
```

a) The "flat" implementation above suffers from serious flaws:
   1. A general numerical algorithm (the Trapezoidal rule) is implemented in a specialized form where the formula for f is inserted directly into the code for the general integration formula.
   2. A general numerical algorithm is not encapsulated as a general function, with appropriate parameters, which can be reused across a wide range of applications.
   3. The lazy programmer dropped the first terms in the general formula since $\sin(0) = \sin(\pi) = 0$.
   4. The sloppy programmer used `np` (number of points?) as variable for `n` in the formula and a counter `k` instead of `i`. Such small deviations from the mathematical notation are completely unnecessary. The closer the code and the mathematics can get, the easier it is to spot errors in formulas.

   Write a function `trapezoidal` that fixes these flaws. Place the function in a module `trapezoidal`.
b) Write a test function `test_trapezoidal`. Call the test function explicitly to check that it works. Remove the call and run pytest on the module:

Terminal

```
Terminal> py.test -s -v trapezoidal
```

Hint Note that even if you know the value of the integral, you do not know the error in the approximation produced by the Trapezoidal rule. However, the Trapezoidal rule will integrate linear functions exactly (i.e., to machine precision). Base a test function on a linear $f(x)$.

c) Add functionality such that we can compute $\int_a^b f(x)dx$ by providing f, a, b, and n as positional command-line arguments to the module file:

Terminal

```
Terminal> python trapezoidal.py 'sin(x)**2' 0 pi 20
```

Here, $a = 0$, $b = \pi$, and $n = 20$.
Note that the `trapezoidal.py` file must still be a valid module file, so the interpretation of command-line data and computation of the integral must be performed from calls in a test block.

Hint To translate a string formula on the command line, like `sin(x)**2`, into a Python function, you can wrap a function declaration around the formula and run `exec` on the string to turn it into live Python code:

```
import math, sys
formula = sys.argv[1]
f_code = """
def f(x):
    return %s
""" % formula
exec(code, math.__dict__)
```

The result is the same as if we had hardcoded

```
from math import *

def f(x):
    return sin(x)**2
```

in the program. Note that `exec` needs the namespace `math.__dict__`, i.e., all names in the `math` module, such that it understands `sin` and other mathematical functions. Similarly, to allow a and b to be `math` expressions like `pi/4` and `exp(4)`, do

Terminal

```
a = eval(sys.argv[2], math.__dict__)
b = eval(sys.argv[2], math.__dict__)
```

d) Write a test function for verifying the implementation of data reading from the command line.

Filename: `trapezoidal`.

Problem 5.3: Implement classes for the Trapezoidal rule

We consider the same problem setting as in Problem 5.2. Make a module with a class `Problem` representing the mathematical problem to be solved and a class `Solver` representing the solution method. The rest of the functionality of the module, including test functions and reading data from the command line, should be as in Problem 5.2.

Filename: `trapezoidal_class`.

Problem 5.4: Write a doctest and a test function

Type in the following program:

```
import sys
# This sqrt(x) returns real if x>0 and complex if x<0
from numpy.lib.scimath import sqrt

def roots(a, b, c):
    """
    Return the roots of the quadratic polynomial
    p(x) = a*x**2 + b*x + c.

    The roots are real or complex objects.
    """
    q = b**2 - 4*a*c
    r1 = (-b + sqrt(q))/(2*a)
    r2 = (-b - sqrt(q))/(2*a)
    return r1, r2

a, b, c = [float(arg) for arg in sys.argv[1:]]
print roots(a, b, c)
```

a) Equip the `roots` function with a doctest. Make sure to test both real and complex roots. Write out numbers in the doctest with 14 digits or less.
b) Make a test function for the `roots` function. Perform the same mathematical tests as in a), but with different programming technology.

Filename: `test_roots`.

Problem 5.5: Investigate the size of tolerances in comparisons

When we replace a comparison `a == b`, where `a` and/or `b` are `float` objects, by a comparison with tolerance, `abs(a-b) < tol`, the appropriate size of `tol` depends on the size of `a` and `b`. Investigate how the size of `abs(a-b)` varies when `b` takes on values 10^k, $k = -5, -9, \ldots, 20$ and `a=1.0/49*b*49`. Thereafter, compute the *relative* difference `abs((a-b)/a)` for the same `b` values.

Filename: `tolerance`.

Remarks You will experience that if `a` and `b` are large, as they can be in, e.g., geophysical applications where lengths measured in meters can be of size 10^6 m, `tol` must be about 10^{-9}, while `a` and `b` around unity can have `tol` of size 10^{-15}. The way out of the problem with choosing a tolerance is to use relative differences.

Exercise 5.6: Make use of a class implementation

Implement the `experiment_compare_dt` function from `decay.py` using class `Problem` and class `Solver` from Sect. 5.5. The parameters `I`, `a`, `T`, the scheme name, and a series of `dt` values should be read from the command line.

Filename: `experiment_compare_dt_class`.

Problem 5.7: Make solid software for a difference equation

We have the following evolutionary difference equation for the number of individuals u^n of a certain specie at time $n\Delta t$:

$$u^{n+1} = u^n + \Delta t\, ru^n \left(1 - \frac{u^n}{M^n}\right), \quad u^0 = U_0 . \tag{5.9}$$

Here, n is a counter in time, Δt is time between time levels n and $n+1$ (assumed constant), r is a net reproduction rate for the specie, and M^n is the upper limit of the population that the environment can sustain at time level n.

Filename: `logistic`.

References

1. W. Gander, M. J. Gander, and F. Kwok, *Scientific Computing - An Introducting Using Maple and MATLAB*. Texts in Computational Science and Engineering. Springer, 2015
2. D. Griffiths, F. David, and D. J. Higham, *Numerical Methods for Ordinary Differential Equations: Initial Value Problems*. Springer, 2010
3. E. Hairer, S. P. Nørsett, and G. Wanner, *Solving Ordinary Differential Equations I. Nonstiff Problems*. Springer, 1993
4. G. Hairer and E. Wanner, *Solving Ordinary Differential Equations II*. Springer, 2010
5. J. D. Hunter, D. Dale, E. Firing, and M. Droettboom, Matplotlib documentation, 2012. http://matplotlib.org/users/
6. H. P. Langtangen, Quick intro to version control systems and project hosting sites. http://hplgit.github.io/teamods/bitgit/html/
7. H. P. Langtangen, SciTools documentation. http://hplgit.github.io/scitools/doc/web/index.html
8. H. P. Langtangen, *A Primer on Scientific Programming with Python*. Texts in Computational Science and Engineering. Springer, fourth edition, 2014
9. H. P. Langtangen and G. K. Pedersen, *Scaling of Differential Equations*. SimulaSpringerBrief. Springer, 2015. http://tinyurl.com/qfjgxmf/web
10. H. P. Langtangen and L. Wang, Odespy software package. https://github.com/hplgit/odespy
11. D. B. Meade and A. A. Struthers, Differential equations in the new millenium: the parachute problem. *International Journal of Engineering Education*, 15(6):417–424, 1999
12. L. Petzold and U. M. Ascher, *Computer Methods for Ordinary Differential Equations and Differential-Algebraic Equations*, volume 61. SIAM, 1998
13. L. N. Trefethen, *Trefethen's index cards - Forty years of notes about People, Words and Mathematics*. World Scientific, 2011

H.P. Langtangen, *Finite Difference Computing with Exponential Decay Models*,
Lecture Notes in Computational Science and Engineering 110,
DOI 10.1007/978-3-319-29439-1

Index

Editorial Policy

1. Volumes in the following three categories will be published in LNCSE:

i) Research monographs
ii) Tutorials
iii) Conference proceedings

Those considering a book which might be suitable for the series are strongly advised to contact the publisher or the series editors at an early stage.

2. Categories i) and ii). Tutorials are lecture notes typically arising via summer schools or similar events, which are used to teach graduate students. These categories will be emphasized by Lecture Notes in Computational Science and Engineering. **Submissions by interdisciplinary teams of authors are encouraged.** The goal is to report new developments – quickly, informally, and in a way that will make them accessible to non-specialists. In the evaluation of submissions timeliness of the work is an important criterion. Texts should be well-rounded, well-written and reasonably self-contained. In most cases the work will contain results of others as well as those of the author(s). In each case the author(s) should provide sufficient motivation, examples, and applications. In this respect, Ph.D. theses will usually be deemed unsuitable for the Lecture Notes series. Proposals for volumes in these categories should be submitted either to one of the series editors or to Springer-Verlag, Heidelberg, and will be refereed. A provisional judgement on the acceptability of a project can be based on partial information about the work: a detailed outline describing the contents of each chapter, the estimated length, a bibliography, and one or two sample chapters – or a first draft. A final decision whether to accept will rest on an evaluation of the completed work which should include

- at least 100 pages of text;
- a table of contents;
- an informative introduction perhaps with some historical remarks which should be accessible to readers unfamiliar with the topic treated;
- a subject index.

3. Category iii). Conference proceedings will be considered for publication provided that they are both of exceptional interest and devoted to a single topic. One (or more) expert participants will act as the scientific editor(s) of the volume. They select the papers which are suitable for inclusion and have them individually refereed as for a journal. Papers not closely related to the central topic are to be excluded. Organizers should contact the Editor for CSE at Springer at the planning stage, see *Addresses* below.

In exceptional cases some other multi-author-volumes may be considered in this category.

4. Only works in English will be considered. For evaluation purposes, manuscripts may be submitted in print or electronic form, in the latter case, preferably as pdf- or zipped ps-files. Authors are requested to use the LaTeX style files available from Springer at http://www.springer.com/gp/authors-editors/book-authors-editors/manuscript-preparation/5636 (Click on LaTeX Template → monographs or contributed books).

For categories ii) and iii) we strongly recommend that all contributions in a volume be written in the same LaTeX version, preferably LaTeX2e. Electronic material can be included if appropriate. Please contact the publisher.

Careful preparation of the manuscripts will help keep production time short besides ensuring satisfactory appearance of the finished book in print and online.

5. The following terms and conditions hold. Categories i), ii) and iii):

Authors receive 50 free copies of their book. No royalty is paid.
Volume editors receive a total of 50 free copies of their volume to be shared with authors, but no royalties.

Authors and volume editors are entitled to a discount of 33.3 % on the price of Springer books purchased for their personal use, if ordering directly from Springer.

6. Springer secures the copyright for each volume.

Addresses:

Timothy J. Barth
NASA Ames Research Center
NAS Division
Moffett Field, CA 94035, USA
barth@nas.nasa.gov

Michael Griebel
Institut für Numerische Simulation
der Universität Bonn
Wegelerstr. 6
53115 Bonn, Germany
griebel@ins.uni-bonn.de

David E. Keyes
Mathematical and Computer Sciences
and Engineering
King Abdullah University of Science
and Technology
P.O. Box 55455
Jeddah 21534, Saudi Arabia
david.keyes@kaust.edu.sa

and

Department of Applied Physics
and Applied Mathematics
Columbia University
500 W. 120 th Street
New York, NY 10027, USA
kd2112@columbia.edu

Risto M. Nieminen
Department of Applied Physics
Aalto University School of Science
and Technology
00076 Aalto, Finland
risto.nieminen@aalto.fi

Dirk Roose
Department of Computer Science
Katholieke Universiteit Leuven
Celestijnenlaan 200A
3001 Leuven-Heverlee, Belgium
dirk.roose@cs.kuleuven.be

Tamar Schlick
Department of Chemistry
and Courant Institute
of Mathematical Sciences
New York University
251 Mercer Street
New York, NY 10012, USA
schlick@nyu.edu

Editor for Computational Science
and Engineering at Springer:
Martin Peters
Springer-Verlag
Mathematics Editorial IV
Tiergartenstrasse 17
69121 Heidelberg, Germany
martin.peters@springer.com

Lecture Notes in Computational Science and Engineering

1. D. Funaro, *Spectral Elements for Transport-Dominated Equations.*
2. H.P. Langtangen, *Computational Partial Differential Equations.* Numerical Methods and Diffpack Programming.
3. W. Hackbusch, G. Wittum (eds.), *Multigrid Methods V.*
4. P. Deuflhard, J. Hermans, B. Leimkuhler, A.E. Mark, S. Reich, R.D. Skeel (eds.), *Computational Molecular Dynamics: Challenges, Methods, Ideas.*
5. D. Kröner, M. Ohlberger, C. Rohde (eds.), *An Introduction to Recent Developments in Theory and Numerics for Conservation Laws.*
6. S. Turek, *Efficient Solvers for Incompressible Flow Problems.* An Algorithmic and Computational Approach.
7. R. von Schwerin, ***M**ulti **B**ody **S**ystem **SIM**ulation.* Numerical Methods, Algorithms, and Software.
8. H.-J. Bungartz, F. Durst, C. Zenger (eds.), *High Performance Scientific and Engineering Computing.*
9. T.J. Barth, H. Deconinck (eds.), *High-Order Methods for Computational Physics.*
10. H.P. Langtangen, A.M. Bruaset, E. Quak (eds.), *Advances in Software Tools for Scientific Computing.*
11. B. Cockburn, G.E. Karniadakis, C.-W. Shu (eds.), *Discontinuous Galerkin Methods.* Theory, Computation and Applications.
12. U. van Rienen, *Numerical Methods in Computational Electrodynamics.* Linear Systems in Practical Applications.
13. B. Engquist, L. Johnsson, M. Hammill, F. Short (eds.), *Simulation and Visualization on the Grid.*
14. E. Dick, K. Riemslagh, J. Vierendeels (eds.), *Multigrid Methods VI.*
15. A. Frommer, T. Lippert, B. Medeke, K. Schilling (eds.), *Numerical Challenges in Lattice Quantum Chromodynamics.*
16. J. Lang, *Adaptive Multilevel Solution of Nonlinear Parabolic PDE Systems.* Theory, Algorithm, and Applications.
17. B.I. Wohlmuth, *Discretization Methods and Iterative Solvers Based on Domain Decomposition.*
18. U. van Rienen, M. Günther, D. Hecht (eds.), *Scientific Computing in Electrical Engineering.*
19. I. Babuška, P.G. Ciarlet, T. Miyoshi (eds.), *Mathematical Modeling and Numerical Simulation in Continuum Mechanics.*
20. T.J. Barth, T. Chan, R. Haimes (eds.), *Multiscale and Multiresolution Methods.* Theory and Applications.
21. M. Breuer, F. Durst, C. Zenger (eds.), *High Performance Scientific and Engineering Computing.*
22. K. Urban, *Wavelets in Numerical Simulation.* Problem Adapted Construction and Applications.
23. L.F. Pavarino, A. Toselli (eds.), *Recent Developments in Domain Decomposition Methods.*
24. T. Schlick, H.H. Gan (eds.), *Computational Methods for Macromolecules: Challenges and Applications.*

25. T.J. Barth, H. Deconinck (eds.), *Error Estimation and Adaptive Discretization Methods in Computational Fluid Dynamics.*

26. M. Griebel, M.A. Schweitzer (eds.), *Meshfree Methods for Partial Differential Equations.*

27. S. Müller, *Adaptive Multiscale Schemes for Conservation Laws.*

28. C. Carstensen, S. Funken, W. Hackbusch, R.H.W. Hoppe, P. Monk (eds.), *Computational Electromagnetics.*

29. M.A. Schweitzer, *A Parallel Multilevel Partition of Unity Method for Elliptic Partial Differential Equations.*

30. T. Biegler, O. Ghattas, M. Heinkenschloss, B. van Bloemen Waanders (eds.), *Large-Scale PDE-Constrained Optimization.*

31. M. Ainsworth, P. Davies, D. Duncan, P. Martin, B. Rynne (eds.), *Topics in Computational Wave Propagation.* Direct and Inverse Problems.

32. H. Emmerich, B. Nestler, M. Schreckenberg (eds.), *Interface and Transport Dynamics.* Computational Modelling.

33. H.P. Langtangen, A. Tveito (eds.), *Advanced Topics in Computational Partial Differential Equations.* Numerical Methods and Diffpack Programming.

34. V. John, *Large Eddy Simulation of Turbulent Incompressible Flows.* Analytical and Numerical Results for a Class of LES Models.

35. E. Bänsch (ed.), *Challenges in Scientific Computing – CISC 2002.*

36. B.N. Khoromskij, G. Wittum, *Numerical Solution of Elliptic Differential Equations by Reduction to the Interface.*

37. A. Iske, *Multiresolution Methods in Scattered Data Modelling.*

38. S.-I. Niculescu, K. Gu (eds.), *Advances in Time-Delay Systems.*

39. S. Attinger, P. Koumoutsakos (eds.), *Multiscale Modelling and Simulation.*

40. R. Kornhuber, R. Hoppe, J. Périaux, O. Pironneau, O. Wildlund, J. Xu (eds.), *Domain Decomposition Methods in Science and Engineering.*

41. T. Plewa, T. Linde, V.G. Weirs (eds.), *Adaptive Mesh Refinement – Theory and Applications.*

42. A. Schmidt, K.G. Siebert, *Design of Adaptive Finite Element Software.* The Finite Element Toolbox ALBERTA.

43. M. Griebel, M.A. Schweitzer (eds.), *Meshfree Methods for Partial Differential Equations II.*

44. B. Engquist, P. Lötstedt, O. Runborg (eds.), *Multiscale Methods in Science and Engineering.*

45. P. Benner, V. Mehrmann, D.C. Sorensen (eds.), *Dimension Reduction of Large-Scale Systems.*

46. D. Kressner, *Numerical Methods for General and Structured Eigenvalue Problems.*

47. A. Boriçi, A. Frommer, B. Joó, A. Kennedy, B. Pendleton (eds.), *QCD and Numerical Analysis III.*

48. F. Graziani (ed.), *Computational Methods in Transport.*

49. B. Leimkuhler, C. Chipot, R. Elber, A. Laaksonen, A. Mark, T. Schlick, C. Schütte, R. Skeel (eds.), *New Algorithms for Macromolecular Simulation.*

50. M. Bücker, G. Corliss, P. Hovland, U. Naumann, B. Norris (eds.), *Automatic Differentiation: Applications, Theory, and Implementations.*

51. A.M. Bruaset, A. Tveito (eds.), *Numerical Solution of Partial Differential Equations on Parallel Computers.*

52. K.H. Hoffmann, A. Meyer (eds.), *Parallel Algorithms and Cluster Computing.*

53. H.-J. Bungartz, M. Schäfer (eds.), *Fluid-Structure Interaction.*

54. J. Behrens, *Adaptive Atmospheric Modeling.*

55. O. Widlund, D. Keyes (eds.), *Domain Decomposition Methods in Science and Engineering XVI.*

56. S. Kassinos, C. Langer, G. Iaccarino, P. Moin (eds.), *Complex Effects in Large Eddy Simulations.*

57. M. Griebel, M.A Schweitzer (eds.), *Meshfree Methods for Partial Differential Equations III.*

58. A.N. Gorban, B. Kégl, D.C. Wunsch, A. Zinovyev (eds.), *Principal Manifolds for Data Visualization and Dimension Reduction.*

59. H. Ammari (ed.), *Modeling and Computations in Electromagnetics: A Volume Dedicated to Jean-Claude Nédélec.*

60. U. Langer, M. Discacciati, D. Keyes, O. Widlund, W. Zulehner (eds.), *Domain Decomposition Methods in Science and Engineering XVII.*

61. T. Mathew, *Domain Decomposition Methods for the Numerical Solution of Partial Differential Equations.*

62. F. Graziani (ed.), *Computational Methods in Transport: Verification and Validation.*

63. M. Bebendorf, *Hierarchical Matrices.* A Means to Efficiently Solve Elliptic Boundary Value Problems.

64. C.H. Bischof, H.M. Bücker, P. Hovland, U. Naumann, J. Utke (eds.), *Advances in Automatic Differentiation.*

65. M. Griebel, M.A. Schweitzer (eds.), *Meshfree Methods for Partial Differential Equations IV.*

66. B. Engquist, P. Lötstedt, O. Runborg (eds.), *Multiscale Modeling and Simulation in Science.*

67. I.H. Tuncer, Ü. Gülcat, D.R. Emerson, K. Matsuno (eds.), *Parallel Computational Fluid Dynamics 2007.*

68. S. Yip, T. Diaz de la Rubia (eds.), *Scientific Modeling and Simulations.*

69. A. Hegarty, N. Kopteva, E. O'Riordan, M. Stynes (eds.), *BAIL* 2008 – *Boundary and Interior Layers.*

70. M. Bercovier, M.J. Gander, R. Kornhuber, O. Widlund (eds.), *Domain Decomposition Methods in Science and Engineering XVIII.*

71. B. Koren, C. Vuik (eds.), *Advanced Computational Methods in Science and Engineering.*

72. M. Peters (ed.), *Computational Fluid Dynamics for Sport Simulation.*

73. H.-J. Bungartz, M. Mehl, M. Schäfer (eds.), *Fluid Structure Interaction II – Modelling, Simulation, Optimization.*

74. D. Tromeur-Dervout, G. Brenner, D.R. Emerson, J. Erhel (eds.), *Parallel Computational Fluid Dynamics 2008.*

75. A.N. Gorban, D. Roose (eds.), *Coping with Complexity: Model Reduction and Data Analysis.*

76. J.S. Hesthaven, E.M. Rønquist (eds.), *Spectral and High Order Methods for Partial Differential Equations.*

77. M. Holtz, *Sparse Grid Quadrature in High Dimensions with Applications in Finance and Insurance.*

78. Y. Huang, R. Kornhuber, O.Widlund, J. Xu (eds.), *Domain Decomposition Methods in Science and Engineering XIX.*

79. M. Griebel, M.A. Schweitzer (eds.), *Meshfree Methods for Partial Differential Equations V.*

80. P.H. Lauritzen, C. Jablonowski, M.A. Taylor, R.D. Nair (eds.), *Numerical Techniques for Global Atmospheric Models.*

81. C. Clavero, J.L. Gracia, F.J. Lisbona (eds.), *BAIL 2010 – Boundary and Interior Layers, Computational and Asymptotic Methods.*

82. B. Engquist, O. Runborg, Y.R. Tsai (eds.), *Numerical Analysis and Multiscale Computations.*

83. I.G. Graham, T.Y. Hou, O. Lakkis, R. Scheichl (eds.), *Numerical Analysis of Multiscale Problems.*

84. A. Logg, K.-A. Mardal, G. Wells (eds.), *Automated Solution of Differential Equations by the Finite Element Method.*

85. J. Blowey, M. Jensen (eds.), *Frontiers in Numerical Analysis – Durham 2010.*

86. O. Kolditz, U.-J. Gorke, H. Shao, W. Wang (eds.), *Thermo-Hydro-Mechanical-Chemical Processes in Fractured Porous Media – Benchmarks and Examples.*

87. S. Forth, P. Hovland, E. Phipps, J. Utke, A. Walther (eds.), *Recent Advances in Algorithmic Differentiation.*

88. J. Garcke, M. Griebel (eds.), *Sparse Grids and Applications.*

89. M. Griebel, M.A. Schweitzer (eds.), *Meshfree Methods for Partial Differential Equations VI.*

90. C. Pechstein, *Finite and Boundary Element Tearing and Interconnecting Solvers for Multiscale Problems.*

91. R. Bank, M. Holst, O. Widlund, J. Xu (eds.), *Domain Decomposition Methods in Science and Engineering XX.*

92. H. Bijl, D. Lucor, S. Mishra, C. Schwab (eds.), *Uncertainty Quantification in Computational Fluid Dynamics.*

93. M. Bader, H.-J. Bungartz, T. Weinzierl (eds.), *Advanced Computing.*

94. M. Ehrhardt, T. Koprucki (eds.), *Advanced Mathematical Models and Numerical Techniques for Multi-Band Effective Mass Approximations.*

95. M. Azaïez, H. El Fekih, J.S. Hesthaven (eds.), *Spectral and High Order Methods for Partial Differential Equations ICOSAHOM 2012.*

96. F. Graziani, M.P. Desjarlais, R. Redmer, S.B. Trickey (eds.), *Frontiers and Challenges in Warm Dense Matter.*

97. J. Garcke, D. Pflüger (eds.), *Sparse Grids and Applications – Munich 2012.*

98. J. Erhel, M. Gander, L. Halpern, G. Pichot, T. Sassi, O. Widlund (eds.), *Domain Decomposition Methods in Science and Engineering XXI.*

99. R. Abgrall, H. Beaugendre, P.M. Congedo, C. Dobrzynski, V. Perrier, M. Ricchiuto (eds.), *High Order Nonlinear Numerical Methods for Evolutionary PDEs – HONOM 2013.*

100. M. Griebel, M.A. Schweitzer (eds.), *Meshfree Methods for Partial Differential Equations VII.*

101. R. Hoppe (ed.), *Optimization with PDE Constraints – OPTPDE 2014.*

102. S. Dahlke, W. Dahmen, M. Griebel, W. Hackbusch, K. Ritter, R. Schneider, C. Schwab, H. Yserentant (eds.), *Extraction of Quantifiable Information from Complex Systems.*

103. A. Abdulle, S. Deparis, D. Kressner, F. Nobile, M. Picasso (eds.), *Numerical Mathematics and Advanced Applications – ENUMATH 2013.*

104. T. Dickopf, M.J. Gander, L. Halpern, R. Krause, L.F. Pavarino (eds.), *Domain Decomposition Methods in Science and Engineering XXII.*

105. M. Mehl, M. Bischoff, M. Schäfer (eds.), *Recent Trends in Computational Engineering – CE2014.* Optimization, Uncertainty, Parallel Algorithms, Coupled and Complex Problems.

106. R.M. Kirby, M. Berzins, J.S. Hesthaven (eds.), *Spectral and High Order Methods for Partial Differential Equations – ICOSAHOM'14.*

107. B. Jüttler, B. Simeon (eds.), *Isogeometric Analysis and Applications 2014.*

108. P. Knobloch (ed.), *Boundary and Interior Layers, Computational and Asymptotic Methods – BAIL 2014.*

109. J. Garcke, D. Pflüger (eds.), *Sparse Grids and Applications – Stuttgart 2014.*

110. H.P. Langtangen, *Finite Difference Computing with Exponential Decay Models.*

For further information on these books please have a look at our mathematics catalogue at the following URL: www.springer.com/series/3527

Monographs in Computational Science and Engineering

1. J. Sundnes, G.T. Lines, X. Cai, B.F. Nielsen, K.-A. Mardal, A. Tveito, *Computing the Electrical Activity in the Heart.*

For further information on this book, please have a look at our mathematics catalogue at the following URL: `www.springer.com/series/7417`

Texts in Computational Science and Engineering

1. H. P. Langtangen, *Computational Partial Differential Equations.* Numerical Methods and Diffpack Programming. 2nd Edition
2. A. Quarteroni, F. Saleri, P. Gervasio, *Scientific Computing with MATLAB and Octave.* 4th Edition
3. H. P. Langtangen, *Python Scripting for Computational Science.* 3rd Edition
4. H. Gardner, G. Manduchi, *Design Patterns for e-Science.*
5. M. Griebel, S. Knapek, G. Zumbusch, *Numerical Simulation in Molecular Dynamics.*
6. H. P. Langtangen, *A Primer on Scientific Programming with Python.* 4th Edition
7. A. Tveito, H. P. Langtangen, B. F. Nielsen, X. Cai, *Elements of Scientific Computing.*
8. B. Gustafsson, *Fundamentals of Scientific Computing.*
9. M. Bader, *Space-Filling Curves.*
10. M. Larson, F. Bengzon, *The Finite Element Method: Theory, Implementation and Applications.*
11. W. Gander, M. Gander, F. Kwok, *Scientific Computing: An Introduction using Maple and MATLAB.*
12. P. Deuflhard, S. Röblitz, *A Guide to Numerical Modelling in Systems Biology.*
13. M. H. Holmes, *Introduction to Scientific Computing and Data Analysis.*
14. S. Linge, H. P. Langtangen, *Programming for Computations* – A Gentle Introduction to Numerical Simulations with MATLAB/Octave.
15. S. Linge, H. P. Langtangen, *Programming for Computations* – A Gentle Introduction to Numerical Simulations with Python.

For further information on these books please have a look at our mathematics catalogue at the following URL: `www.springer.com/series/5151`

Printed in the United States
By Bookmasters